METHODS IN MOLECULAR BIOLOGY

Series Editor
John M. Walker
School of Life and Medical Sciences
University of Hertfordshire
Hatfield, Hertfordshire, UK

For further volumes:
http://www.springer.com/series/7651

For over 35 years, biological scientists have come to rely on the research protocols and methodologies in the critically acclaimed *Methods in Molecular Biology* series. The series was the first to introduce the step-by-step protocols approach that has become the standard in all biomedical protocol publishing. Each protocol is provided in readily-reproducible step-by-step fashion, opening with an introductory overview, a list of the materials and reagents needed to complete the experiment, and followed by a detailed procedure that is supported with a helpful notes section offering tips and tricks of the trade as well as troubleshooting advice. These hallmark features were introduced by series editor Dr. John Walker and constitute the key ingredient in each and every volume of the *Methods in Molecular Biology* series. Tested and trusted, comprehensive and reliable, all protocols from the series are indexed in PubMed.

Laser Capture Microdissection

Methods and Protocols

Edited by

Roberta Noberini

Department of Experimental Oncology, European Institute of Oncology IRCCS (IEO), Milano, Italy

Humana Press

Editor
Roberta Noberini
Department of Experimental Oncology
European Institute of Oncology IRCCS (IEO)
Milano, Italy

ISSN 1064-3745 ISSN 1940-6029 (electronic)
Methods in Molecular Biology
ISBN 978-1-0716-5153-7 ISBN 978-1-0716-5154-4 (eBook)
https://doi.org/10.1007/978-1-0716-5154-4

This Humana imprint is published by the registered company Springer Science+Business Media, LLC, part of Springer Nature.
The registered company address is: 1 New York Plaza, New York, NY 10004, U.S.A.

Preface

Laser capture microdissection (LCM) is a microscope-based technique that enables the isolation of subpopulations of cells from tissue and other specimens under direct microscopic visualization. Since its first application in the early 1920s, LCM has evolved into a powerful tool used across diverse fields of biomedical research, including cancer, neuroscience, developmental biology, and forensic science.

LCM allows the selective isolation and collection of homogeneous cell populations, eliminating unwanted cells from complex tissues. Thanks to recent advances in optical resolution and instrumentation, LCM also enables the collection of individual cells for single-cell downstream applications. While it is traditionally applied to mammalian tissues—both formalin-fixed paraffin-embedded (FFPE) and cryosectioned—LCM is also suitable for a variety of other sample types, including blood smears and live cells, which can be isolated and collected for propagation.

This volume brings together a comprehensive set of protocols that highlight the versatility of LCM across different biological contexts and analytical workflows. The chapters cover protocols for isolating cells from blood smears, live cell samples, and FFPE or frozen tissues. Applications range from lipidomics, transcriptomics (e.g., RNA-seq, Geo--seq), epigenomics (e.g., DNA methylation profiling, ATAC-seq, ChIP-seq), and spatial glycomics to a wide array of proteomic approaches, both for the analysis of whole proteomes and specific protein categories (histones and amyloid proteins). Also included are emerging AI-driven applications of LCM, such as artificial intelligence-guided and automated microscopy for high-precision, spatially resolved cell selection.

The collection represents an up-to-date and practical resource for researchers at the interface of microscopy, cell biology, and molecular profiling. By providing detailed, step-by-step methods, this volume is designed to support both newcomers and experienced users in effectively applying LCM to their own biological questions, across a wide variety of systems and technical platforms.

Milano, Italy — ***Roberta Noberini***

Contents

Contributors

JENNIFER AGUILAN • *Department of Pathology, Albert Einstein College of Medicine, Bronx, NY, USA*

MARC D. BEYER • *Immunogenomics & Neurodegeneration, Deutsches Zentrum für Neurodegenerative Erkrankungen (DZNE) e.V, Bonn, Germany; Systems Medicine, Deutsches Zentrum für Neurodegenerative Erkrankungen (DZNE) e.V, Bonn, Germany; PRECISE Platform for Genomics and Epigenomics, Deutsches Zentrum für Neurodegenerative Erkrankungen (DZNE) e.V. and University of Bonn and West German Genome Center, Bonn, Germany*

TIZIANA BONALDI • *Department of Experimental Oncology, IEO, European Institute of Oncology IRCCS, Milan, Italy; Department of Oncology and Haematology-Oncology, University of Milano, Milan, Italy*

PATCHARAPORN BOOTTANUN • *Cellular and Molecular Biotechnology Research Institute, National Institute of Advanced Industrial Science and Technology (AIST), Tsukuba, Ibaraki, Japan; Glycan and Life Systems Integration Center (GaLSIC), Soka University, Hachioji, Tokyo, Japan*

LUCA CALABRESE • *Department of Otolaryngology, Provincial Hospital of Bolzano (SABES-ASDAA), Lehrkrankenhaus der Paracelsus Medizinischen Privatuniversität, Bolzano, Italy; Division of Otorhinolaryngology, "San Maurizio" Hospital, Bolzano, Italy*

CATERINA CARRARO • *Systems Medicine, Deutsches Zentrum für Neurodegenerative Erkrankungen (DZNE) e.V, Bonn, Germany; Genomics and Immunoregulation, Life and Medical Sciences (LIMES) Institute, University of Bonn, Bonn, Germany*

YINGYING CHEN • *Guangzhou National Laboratory, Guangzhou International Bio Island, Guangzhou, Guangdong, China*

FABIAN COSCIA • *Max-Delbrück-Center for Molecular Medicine in the Helmholtz Association (MDC), Spatial Proteomics Group, Berlin, Germany*

ALESSANDRO CUOMO • *European Institute of Oncology (IEO), Mass Spectrometry Unit (MSU), Milan, Italy*

PIERLUIGI DI CHIARO • *Laboratory of Translational Research, Azienda Unità Sanitaria Locale-IRCCS di Reggio Emilia, Reggio Emilia, Italy*

GIUSEPPE R. DIAFERIA • *Botton-Champalimaud Pancreatic Cancer Center, Champalimaud Foundation, Lisbon, Portugal*

EKATERINA A. ELESINA • *Shemyakin-Ovchinnikov Institute of Bioorganic Chemistry, Russian Academy of Sciences, Moscow, Russia; Lomonosov Moscow State University, Moscow, Russia*

MARIO FARETTA • *Department of Experimental Oncology, European Institute of Oncology IRCCS, Milan, Italy*

SONJA FRITZSCHE • *Max-Delbrück-Center for Molecular Medicine in the Helmholtz Association (MDC), Spatial Proteomics Group, Berlin, Germany; Humboldt-Universität zu Berlin, Institute of Biology, Berlin, Germany*

LAURA FURIA • *Department of Experimental Oncology, European Institute of Oncology IRCCS, Milan, Italy*

SAYAKA FUSEYA • *Cellular and Molecular Biotechnology Research Institute, National Institute of Advanced Industrial Science and Technology (AIST), Tsukuba, Ibaraki, Japan*

TOMMASO GASPARELLO • *Institute of Genetics and Cancer, University of Edinburgh, Edinburgh, UK*

ILARIA GIROLAMI • *Department of Pathology, Provincial Hospital of Bolzano (SABES-ASDAA), Lehrkrankenhaus der Paracelsus Medizinischen Privatuniversität, Bolzano, Italy*

ROBERTO GRIGOLATO • *Department of Biomedical, Surgical and Dental Sciences, University of Milan, Milan, Italy*

EMI HANAWA • *Department of Genome Biology, Institute of Medicine, University of Tsukuba, Ibaraki, Japan*

CELIA L. HARTMANN • *Immunogenomics & Neurodegeneration, Deutsches Zentrum für Neurodegenerative Erkrankungen (DZNE) e.V, Bonn, Germany; Systems Medicine, Deutsches Zentrum für Neurodegenerative Erkrankungen (DZNE) e.V, Bonn, Germany*

LENA HERSEMANN • *Max Planck Institute of Molecular Cell Biology and Genetics, Dresden, Germany*

MAXWELL HORTON • *Department of Biochemistry, Albert Einstein College of Medicine, Bronx, NY, USA; Department of Molecular Pharmacology, Albert Einstein College of Medicine, Bronx, NY, USA*

YINGCHUO HU • *Changping Laboratory, Beijing, China; Academy for Advanced Interdisciplinary Studies, Peking University, Beijing, China; Biomedical Pioneering Innovation Center (BIOPIC), Peking University, Beijing, China*

YANYI HUANG • *Biomedical Pioneering Innovation Center (BIOPIC), Peking University, Beijing, China; Beijing National Laboratory for Molecular Sciences, College of Chemistry and Molecular Engineering, Peking University, Beijing, China; Institute for Cell Analysis, Shenzhen Bay Laboratory, Shenzhen, China*

NAIHE JING • *Guangzhou National Laboratory, Guangzhou International Bio Island, Guangzhou, Guangdong, China; Guangzhou Medical University, Guangzhou, Guangdong, China*

NAOMI KANEKO • *Department of Genome Biology, Institute of Medicine, University of Tsukuba, Ibaraki, Japan*

ULYANA V. KHLEBNIKOVA • *Shemyakin-Ovchinnikov Institute of Bioorganic Chemistry, Russian Academy of Sciences, Moscow, Russia; Lomonosov Moscow State University, Moscow, Russia*

ATSUSHI KUNO • *Cellular and Molecular Biotechnology Research Institute, National Institute of Advanced Industrial Science and Technology (AIST), Tsukuba, Ibaraki, Japan*

LU LIU • *Changping Laboratory, Beijing, China*

HIDEMASA MATSUO • *Human Health Sciences, Graduate School of Medicine, Kyoto University, Kyoto, Japan*

HONGKEE MOON • *Max Planck Institute of Molecular Cell Biology and Genetics, Dresden, Germany*

MASAFUMI MURATANI • *Department of Genome Biology, Institute of Medicine, University of Tsukuba, Ibaraki, Japan; Transborder Medical Research Center, University of Tsukuba, Tsukuba, Ibaraki, Japan*

DANIELE MUSIANI • *European Institute of Oncology (IEO), Mass Spectrometry Unit (MSU), Milan, Italy*

LUCIA NACCI • *Department of Experimental Oncology, IEO, European Institute of Oncology IRCCS, Milano, Italy*

GAYATHRI NADAR • *Discovery Life Sciences Biomarker Services GmbH, Kassel, Germany*

Chiaki Nagai-Okatani • *Cellular and Molecular Biotechnology Research Institute, National Institute of Advanced Industrial Science and Technology (AIST), Tsukuba, Ibaraki, Japan*

Roberta Noberini • *Department of Experimental Oncology, European Institute of Oncology IRCCS (IEO), Milano, Italy*

Masayuki Noguchi • *Center for Clinical and Translational Science, Shonan Kamakura General Hospital, Kamakura, Japan; Department of Pathology, Narita Tomisato Tokushukai Hospital, Chiba, Japan*

Matthew Pearson • *Institute of Genetics and Cancer, University of Edinburgh, Edinburgh, UK*

Simone Pelicci • *Department of Experimental Oncology, European Institute of Oncology IRCCS, Milan, Italy*

Jeffrey E. Pessin • *Department of Molecular Pharmacology, Albert Einstein College of Medicine, Bronx, NY, USA; Department of Medicine, Albert Einstein College of Medicine, Bronx, NY, USA*

Oleg V. Podgorny • *Shemyakin-Ovchinnikov Institute of Bioorganic Chemistry, Russian Academy of Sciences, Moscow, Russia; Pirogov Russian National Research Medical University, Moscow, Russia; Federal Center of Brain Research and Neurotechnologies, Federal Medical Biological Agency, Moscow, Russia*

Silvia Radrezza • *Max Planck Institute of Molecular Cell Biology and Genetics, Dresden, Germany*

Joachim L. Schultze • *Systems Medicine, Deutsches Zentrum für Neurodegenerative Erkrankungen (DZNE) e.V, Bonn, Germany; PRECISE Platform for Genomics and Epigenomics, Deutsches Zentrum für Neurodegenerative Erkrankungen (DZNE) e.V. and University of Bonn and West German Genome Center, Bonn, Germany; Genomics and Immunoregulation, Life and Medical Sciences (LIMES) Institute, University of Bonn, Bonn, Germany*

Ivey Sebastian • *Max Planck Institute of Molecular Cell Biology and Genetics, Dresden, Germany*

Andrej Shevchenko • *Max Planck Institute of Molecular Cell Biology and Genetics, Dresden, Germany*

Simone Sidoli • *Department of Biochemistry, Albert Einstein College of Medicine, Bronx, NY, USA*

Francesco Spadari • *Department of Biomedical, Surgical and Dental Sciences, University of Milan, Milan, Italy; Maxillofacial Surgery and Dental Unit, Fondazione IRCCS Cà Granda Ospedale Maggiore Policlinico, Milan, Italy*

Alex von Kriegsheim • *Institute of Genetics and Cancer, University of Edinburgh, Edinburgh, UK*

Xianfa Yang • *Zhejiang University-University of Edinburgh Institute (ZJU-UoE Institute), Zhejiang University School of Medicine, Haining, China; Guangzhou National Laboratory, Guangzhou International Bio Island, Guangzhou, Guangdong, China*

Yun Yang • *Guangzhou National Laboratory, Guangzhou International Bio Island, Guangzhou, Guangdong, China; Guangzhou Medical University, Guangzhou, Guangdong, China*

Chunhong Zheng • *Frontiers Science Center for Cancer Integrative Omics, Peking University International Cancer Institute, Peking University, Beijing, China*

Chapter 1

Laser Capture Microdissection from Blood Smears

Hidemasa Matsuo

Abstract

The technology of laser capture microdissection (LCM) is used to isolate specific regions of tissue samples or specific cells under microscopic observation. Several studies have employed LCM to collect tissue samples or cells from pathological tissues. It can also be used to collect cells from blood smears, particularly when it is difficult to differentiate cells based on expression markers or when the sample volume is limited. This chapter describes a method for preparing peripheral blood or bone marrow smears on glass slides with membranes, followed by staining with May-Grünwald Giemsa solution and cell collection via LCM. In addition, methods for extracting DNA and RNA from the collected cells for the analysis of gene aberrations and expression levels are described.

Key words Laser capture microdissection, Blood smears, Peripheral blood, Bone marrow, Leukemia

1 Introduction

Laser capture microdissection (LCM) is used to isolate specific regions of tissue samples or specific cells under microscopic observation [1, 2]. The two main types of LCM systems are infrared (IR) capture systems and ultraviolet (UV) cutting systems [3]. In both systems, the first step involves the microscopic visualization of the target cells. Subsequently, cell collection is enabled by the transfer of laser energy to a thermolabile polymer with the formation of a polymer-cell composite. In contrast, a UV cutting system involves photoablation of an area around the cell by a UV laser, thereby enabling cell collection. Several LCM instruments are currently available, including models that combine IR and UV technologies into a single device [4].

The advantage of LCM is that based on morphological information, it enables the collection of highly pure tissues and cells from heterogeneous cell populations. In addition, it is compatible with various tissue/cell types and cellular staining methods and facilitates downstream genomic, transcriptomic, and proteomic

Roberta Noberini (ed.), *Laser Capture Microdissection: Methods and Protocols*, Methods in Molecular Biology, vol. 3015,
https://doi.org/10.1007/978-1-0716-5154-4_1,

analyses [5–8]. Although most studies use LCM to collect regions of interest from pathological tissues, it also enables the collection of cells from blood smears [9]. The bone marrow and peripheral blood cells are heterogeneous populations composed of cells belonging to various differentiation stages and lineages. Blood cell identification is commonly performed using methods such as fluorescence-activated cell sorting (FACS), which utilizes markers that are characteristically expressed in the normal/abnormal cell population [10, 11]. However, identifying cells based on expression markers alone is difficult because the immunophenotype of blood cells is not always consistent across patients [12]. Moreover, while cell loss is generally unavoidable during FACS sample preparation and sorting, LCM offers the advantage of enabling, in principle, the collection of all cells present on a membrane slide, thus allowing analysis from even a small specimen. Failure to obtain bone marrow on attempted marrow aspiration is referred to as "dry tap" and has been reported in 3%–10% of bone marrow aspirations [13]. Such cases require methods that analyze minute amounts of samples. Therefore, LCM represents a valuable tool for improving our understanding of hematologic diseases.

I hereby present a method for preparing smears of peripheral blood or bone marrow on glass slides using a membrane, staining the cells with May-Grünwald Giemsa solution, and collecting cells by LCM. This section also presents methods for extracting DNA and RNA to analyze gene aberrations and expression levels in collected cells.

2 Materials

2.1 Blood Smear Preparation for LCM

1. Bone marrow specimen or peripheral blood. Typically, 10 μL of sample is used per slide, with a minimum requirement of approximately 5 μL.
2. Polyethylene naphthalate (PEN) membrane-coated glass slides. This method uses RNase-free membrane slides 2.0 μm (PEN) (Leica, cat no. 11505273).
3. Spreader slide (e.g., cover glass).
4. Electric fan.

2.2 May-Grünwald Giemsa Staining

1. May-Grünwald stain solution.
2. 0.0667 mol/L phosphate buffer solution (pH 6.4). Milli-Q-purified water is used to prepare a 1:10 dilution.
3. Giemsa stain solution.
4. Milli-Q-purified water for washing purposes.

2.3 LCM

1. LCM instrument. LMD7000 (Leica) is used in this method.
2. Collection tubes. The selection of the collection tube depends on the LCM instrument used. A 0.2 mL PCR (polymerase chain reaction) tube is used in this protocol (Greiner, cat. no. 683201).
3. Lysis buffer for cell collection. The choice of the lysis buffer depends on the downstream assays. The appropriate type/amount of lysis buffer required is 20 μL of Arcturus® PicoPure® protease buffer provided with the Arcturus® PicoPure® DNA Extraction Kit for DNA extraction or 75 μL of Buffer RLT provided by RNeasy® Micro Kit for RNA extraction.

2.4 DNA and RNA Extraction

1. DNA extraction kit. The Arcturus® PicoPure® DNA Extraction Kit (ThermoFisher) is used in this protocol.
2. RNA extraction kit. The RNeasy® Micro Kit (QIAGEN) is used in this method.
3. Benchtop centrifuge.
4. Thermal cycler (for DNA extraction).

3 Methods

All procedures should be performed at room temperature unless otherwise specified.

3.1 Preparation of Blood Smears for LCM

1. Place a small drop of peripheral blood/bone marrow (typically around 10 μL; a minimum of approximately 5 μL per slide is required) onto PEN membrane-coated glass slides.
2. Prepare the blood smear by gently moving a spreader slide in a manner different from conventional smear techniques to avoid damaging the membrane (*see* Fig. 1a, b and **Note 1**).
3. Dry the smears rapidly and thoroughly under an electric fan (*see* **Note 2**).
4. After fixing and staining the smears with May-Grünwald stain solution for 5 minutes, gently wash the smears with Milli-Q-purified water (*see* **Note 3**).
5. Prepare the Giemsa staining solution by mixing 20 mL of 1/150 mol/L phosphate buffer solution with 1.5 mL of Giemsa stain solution. This volume is sufficient to stain approximately five smears.
6. Stain the smears with the prepared Giemsa solution for 20 minutes, followed by a gentle rinse with Milli-Q-purified water, and allow the smears to dry (*see* **Note 4**).

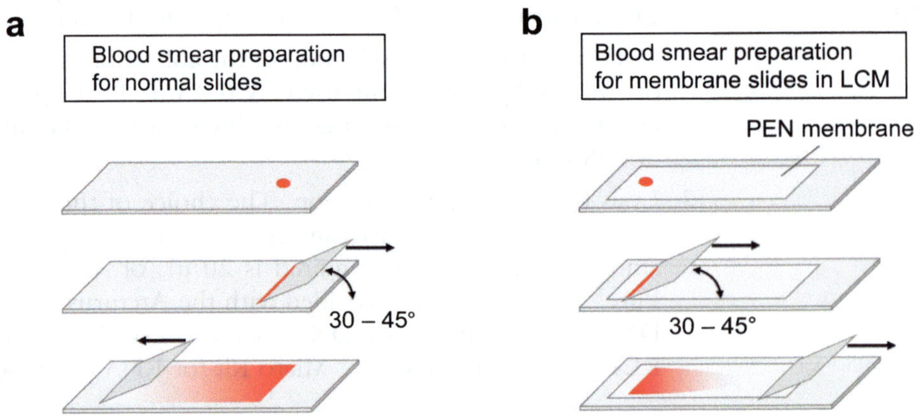

Fig. 1 Methods of preparing the blood smears on the membrane slides for LCM compared with normal slide preparation. (**a**) For standard smear preparation, a small drop of peripheral blood/bone marrow is first deposited on a glass slide. A spreader slide is positioned at an angle of 30°–45° in front of the drop and then drawn backward to make contact with it. The blood spreads along the line of contact, and a smooth forward of the spreader slide is used to create the smear. (**b**) For smear preparation on PEN membrane slides intended for LCM, the spreader slide must be moved gently in the opposite direction to avoid damaging the membrane. The spreader slide should not be pushed on the bottom slide

7. The smears have to be stored at appropriate temperatures for their intended use (for DNA analysis, room temperature or 4 °C; for RNA analysis, −80 °C) (*see* **Note 5**).

3.2 Cell Collection by LCM

1. Turn on the LCM instrument and calibrate it in accordance with the manufacturer's instructions.
2. Set the slide with the membrane facing down.
3. Set a collection tube to the collector with a lever for the tube caps of the LCM instruments. Add the appropriate type and amount of lysis buffer to the lids of the collection tubes (*see* Fig. 2a).
4. Initially observe the cells at low magnification to bring them into focus and then observe them at high magnification to select the cells to be collected.
5. Set the values for each parameter in the laser control. Examples of values (magnification: 40×) for collecting blood cells from smears using LMD7000 are given in Table 1 (*see* **Note 6**).
6. Draw a line around the cells to be collected and irradiate them with a UV laser for cell collection (*see* Fig. 2b and **Note 7**).
7. Collect the target number of cells. According to the experience of the author's group, approximately 100 cells, preferably 300 cells, are required for reproducible mutation analysis and gene expression level results (*see* **Note 8**).

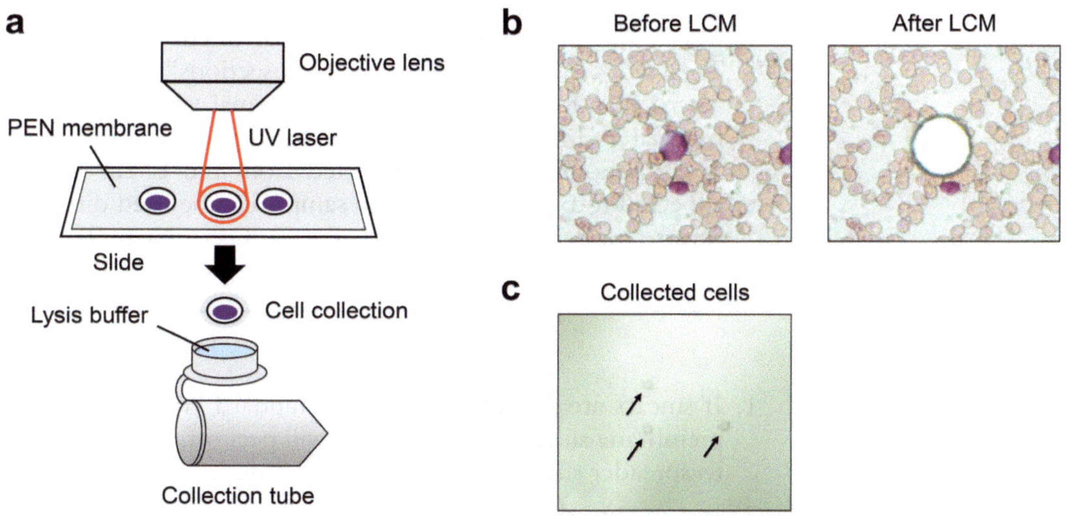

Fig. 2 Collection of cells from blood smears (**a**) A schematic figure illustrating cell collection from blood smears using the LMD7000. (**b**) Microscopic images of blood smears before and after the LCM. (**c**) Collected cells in the lid of the collection tube (arrows)

Table 1
Examples of the parameter values for the laser control in the LMD7000 (magnification: 40×)

Parameters	Value
Power	9
Aperture	3
Speed	10
Specimen balance	1
Head current (%)	70
Pulse frequency	119

8. Confirm the presence of the cells in the lid of the collection tube at the time of collection (*see* Fig. 2c).
9. Close the tube lid carefully and spin it down.

3.3 Isolation of Nucleic Acids

1. For DNA extraction, follow the manufacturer's instructions provided with the Arcturus® PicoPure® DNA Extraction Kit. Additional guidance can be found in a previous study that utilized this kit [14]. Briefly, the procedure is as follows: Lightly vortex the tube for 5–10 seconds, centrifuge at 1500 g for 1 minute at 4 °C, then place it on the thermal cycler and run the following program: 60 °C for 3 hours, 75 °C for 30 minutes, and hold at 4 °C. The sample can be used directly for PCR or stored at −20 °C.

2. RNA extraction should be performed using the RNeasy® Micro Kit, according to the protocol titled "Purification of Total RNA from Microdissected Cryosections" in the RNeasy® Micro Handbook (QIAGEN). A minimal volume of RNase-free water (14 μL) should be used for the final elution of RNA as the amount of RNA extracted from the LMD sample is expected to be very small. The sample can be used directly for reverse transcription or stored at −80 °C (*see* **Note 9**).

4 Notes

1. If smears are prepared using the standard Wedge method on membrane slides, the membrane will peel off. As an alternative to spreader slides, pipette tips can be used.
2. Incomplete drying can cause blood cells to detach from the membrane during the May-Grünwald Giemsa staining process.
3. Careful and gentle washing is vital for preventing the detachment of the blood cells from the membrane.
4. If blood cells are insufficiently stained, the staining time can be extended to 30 minutes. If necessary, wipe off the staining solution on the underside of the membrane slides with alcohol to make them suitable for observation.
5. If the membrane slides are stored at −80 °C for a long time, frost will form on the surfaces of slides not stored in airtight containers. To avoid this, it is advisable to store the slides individually in 50 mL centrifuge tubes and seal the tubes with Parafilm®.
6. The described laser settings are applicable to peripheral blood and bone marrow smears. However, it may be necessary to increase the laser power or to irradiate the laser multiple times in areas of high cell density to collect the cells.
7. With regard to leukocytes, collection should be avoided to prevent contamination if the cells of interest are in close proximity to other nucleated cells.
8. The number of cells to be collected by LCM depends on the purpose of the experiment and should be determined after referring to similar studies. A study in the literature summarizes the use of LCM for various purposes [2].
9. Based on the author's experience, it is sometimes difficult to analyze gene expression levels from the LCM samples once the RNA sample has been stored. Therefore, it is recommended that RNA extracted from LCM samples should not be stored and that downstream assays such as reverse transcription should be performed.

Acknowledgments

This work was supported by a Grant-in-Aid from JSPS KAKENHI (JP22K08475 and JP25K02623). The author is grateful to Dr. Masao Doi and Mr. Yuki Hamada (Department of Systems Biology, Graduate School of Pharmaceutical Sciences, Kyoto University, Kyoto, Japan) for providing access to the LMD7000. Although the results of the genomic analysis are not included in this chapter, the author thanks Dr. Kenichi Yoshida (Division of Cancer Evolution, National Cancer Center Research Institute, Tokyo, Japan) for his support in conducting the genomic analysis of the LCM samples, which helped the author obtain useful information for writing this method.

References

1. Emmert-Buck MR, Bonner RF, Smith PD et al (1996) Laser capture microdissection. Science 274:998–1001
2. Espina V, Wulfkuhle JD, Calvert VS et al (2006) Laser-capture microdissection. Nat Protoc 1:586–603
3. Datta S, Malhotra L, Dickerson R (2015) Laser capture microdissection: big data from small samples. Histol Histopathol 30:1255–1269
4. Gallagher RI, Blakely SR, Liotta LA et al (2012) Laser capture microdissection: Arcturus(XT) infrared capture and UV cutting methods. Methods Mol Biol 823:157–178
5. Xue Y, Liu L, Zhang Y et al (2024) Unraveling the key role of chromatin structure in cancer development through epigenetic landscape characterization of oral cancer. Mol Cancer 23:190
6. Nishikubo M, Matsuo H, Manabe S et al (2024) Donor-derived CD8-predominant T cells in human graft-versus-host disease of the brain after allogeneic transplantation. Br J Haematol 204:2516–2519
7. Göcz B, Rumpler É, Szentkirályi-Tóth S et al (2025) Laser-capture microdissection for spatial transcriptomics of immunohistochemically detected neurons. J Biol Chem 301:108150
8. Ezzoukhry Z, Henriet E, Cordelières FP et al (2018) Combining laser capture microdissection and proteomics reveals an active translation machinery controlling invadosome formation. Nat Commun 9:2031
9. Matsuo H, Shiga S, Imai T et al (2017) Purification of leukemic blast cells from blood smears using laser microdissection. Int J Hematol 106:55–59
10. Anguille S, Van Tendeloo VF, Berneman ZN (2012) Leukemia-associated antigens and their relevance to the immunotherapy of acute myeloid leukemia. Leukemia 26:2186–2196
11. Döhner H, Estey E, Grimwade D et al (2017) Diagnosis and management of AML in adults: 2017 ELN recommendations from an international expert panel. Blood 129:424–447
12. Matsuo H, Wakita T, Hiramatsu H et al (2021) Blast cells in acute megakaryoblastic leukaemia with Down syndrome are characterized by low CLEC12A expression. Br J Haematol 192:e7–e11
13. Donald S, Kakkar N (2021) Dry tap on bone marrow aspiration: a red flag. J Hematop 14:125–130
14. Ellis P, Moore L, Sanders MA et al (2021) Reliable detection of somatic mutations in solid tissues by laser-capture microdissection and low-input DNA sequencing. Nat Protoc 16:841–871

Chapter 2

Live Cell Isolation by Laser Capture Microdissection

Ekaterina A. Elesina, Ulyana V. Khlebnikova, and Oleg V. Podgorny

Abstract

Molecular profiling is a powerful strategy for dissecting molecular aspects of cell functioning. The precision of molecular profiling is highly dependent on purity and homogeneity of cell samples. Cell technologies need cell populations enriched with target cells. Numerous approaches have been proposed to isolate and collect homogenous populations of cells and even single cells for downstream molecular profiling and recultivation. Unlike other methods for isolation of live cells, laser microdissection allows for collecting single cells and cell colonies from adherent cell cultures without detaching cells from the substrate. This is advantageous for subsequent downstream omics analyses or recultivation because of preserving cells in their intact state. Here, we present a protocol for live cell laser microdissection with gravity transfer. This approach allows for isolating fragments of cell monolayers and single cells for downstream molecular profiling and recultivation.

Key words Laser microdissection, Cell culture, Recultivation, Cell separation, Isolation of live cells

1 Introduction

Laser microdissection is a technique that is used for the isolation of cells, tissue fragments, or cellular components for various downstream analyses, including bulk and single-cell omics technologies [1–7].

The underlying principle of laser microdissection methods is cold ablation, also known as plasma-induced ablation [8, 9]. During this process, material is dissociated by high-intensity ultraviolet (UV) radiation without thermal energy transfer to the surrounding environment. When UV-absorbing material (for instance, polyethylene naphthalate [PEN] foil) is irradiated by a high-energy laser at the focal point, energy absorption breaks molecular bonds, ionizes the material, and generates microplasma. The excessive energy stored within microplasma drives the process of its expansion. Consequently, material dissociation occurs without heat transfer to the environment, which makes it applicable for live cell microdissection [10].

Roberta Noberini (ed.), *Laser Capture Microdissection: Methods and Protocols*, Methods in Molecular Biology, vol. 3015,
https://doi.org/10.1007/978-1-0716-5154-4_2,

The major advantage of live cell laser capture microdissection over other existing techniques is that it provides isolation and collection of cell colonies and even single cells from adherent cell cultures without their detachment from the substrate. Detaching adherent cells from the substrates usually requires treatment with mixtures of enzymes digesting components of the cellular surface and extracellular matrix. Such treatment induces cell responses that can affect downstream molecular profiling and elicit the loss of cell identity after recultivation. In contrast to other live cell isolation techniques, live cell laser capture microdissection preserves the intact state of collected cells.

The current protocol describes a laser microdissection with gravity transfer (LMDGT) procedure for the successful recultivation of live cells and colony fragments. The scheme of the procedure is shown in Fig. 1. The Leica laser microdissection systems are mounted on an upright microscope and use gravity for sample collection. This enables automated sample transfer to the collector to be carried out without direct operator intervention.

This protocol allows for recultivating fragments of the HeLa monolayer. The viability of cells present within dissected samples was found to be greater than 50%, as demonstrated in Fig. 2. Samples with a diameter greater than 200 μm showed 100% recultivation efficiency (Fig. 3). LMDGT allows for single-cell cloning to be carried out (Fig. 4). The method described here has been successfully used for Chlamydia-infected cells isolation and recultivation for whole genome sequencing [11]. Furthermore, there are some studies that have employed this method for collecting the material for RNA sequencing [12] and qRT-PCR [13].

2 Materials

2.1 Cell Culturing

1. HeLa Kyoto (EMBL collection, gender: F, RRID: CVCL_1922).
2. Culture medium: DMEM with 4.5 g/L D-glucose, 10% fetal bovine serum, 100 μg/mL streptomycin, 100 U/mL penicillin (*see* **Note 1**).
3. Petri dish with PEN (polyethylene naphthalate) membrane (Leica Microsystems, dish diameter: 50 mm, membrane thickness: 2 μm).
4. 0.01% poly-D-lysine solution, MW: 70000–150,000.
5. Hanks' Balanced Salt Solution (HBSS) with 10 mM HEPES-Na.
6. 96-well strip plates for samples recultivation.
7. Laminar flow cabinet.
8. Humidified incubator (37 °C, 5% CO_2).

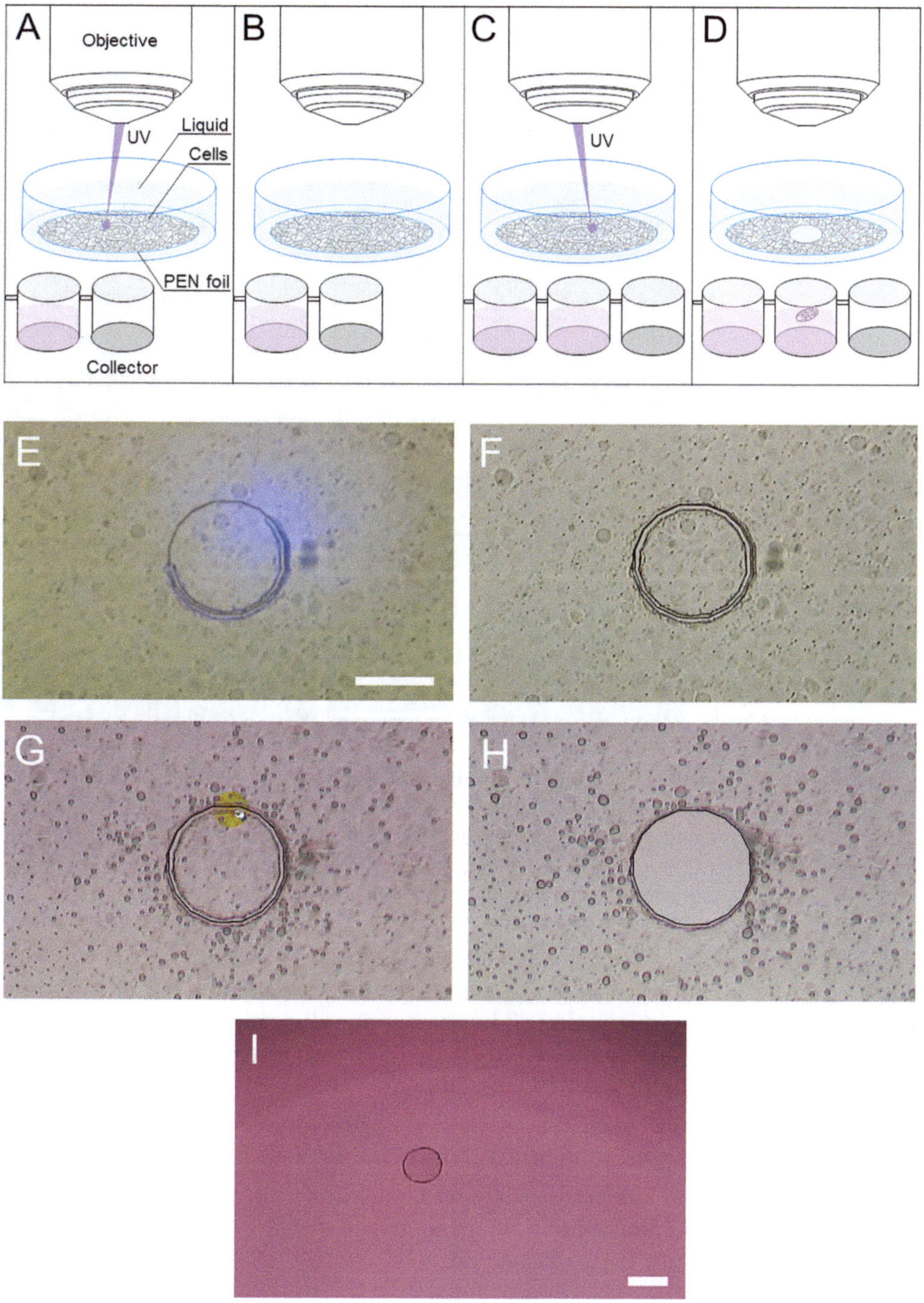

Fig. 1 Scheme of the laser microdissection with gravity transfer. Panels (**a**–**d**) illustrate the live cell laser microdissection workflow. After isolating a circular sample containing cells, a laser pulse is applied to the edge of the circle, shooting the specimen down into the collector. Microphotographs (**e**–**i**) illustrate the process of live cell isolation. Scale bar (**e**–**h**): 100 μm. Scale bar (**i**): 200 μm

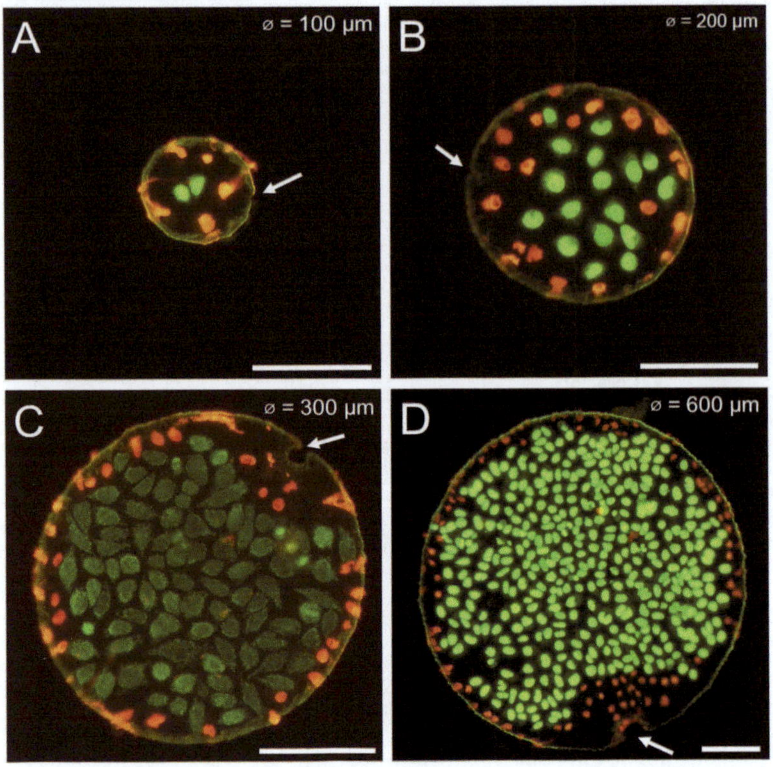

Fig. 2 Distribution of viable and dead cells on samples of different diameters after laser microdissection. The samples were stained with SYBR Green (green) and propidium iodide (red). Sample diameters were 100 μm (**a**), 200 μm (**b**), 300 μm (**c**), and 600 μm (**d**). Arrows show the points of the laser pulse application for extracting the sample. Scale bars: 100 μm

2.2 Laser Microdissection Instrumentation

1. Leica LMD7 laser microdissection system (Leica Microsystems, Germany). The laser microdissection system should have a climatic chamber for temperature control during the isolation procedure.
2. Nd:YLF laser ($\lambda = 349$ nm, pulse duration: 4 ns, maximum pulse energy:120 μJ, tunable pulse frequency: 10–5000 Hz).
3. Objectives: 10× with the numerical aperture of 0.3, 20× with the numerical aperture of 0.4.

3 Methods

3.1 Cell Cultivation

1. Grow HeLa Kyoto cells in 25 cm^2 flasks with culture medium DMEM at 37 °C with 5% CO_2. Cell cultivation and dissection procedures should be carried out under sterile conditions.

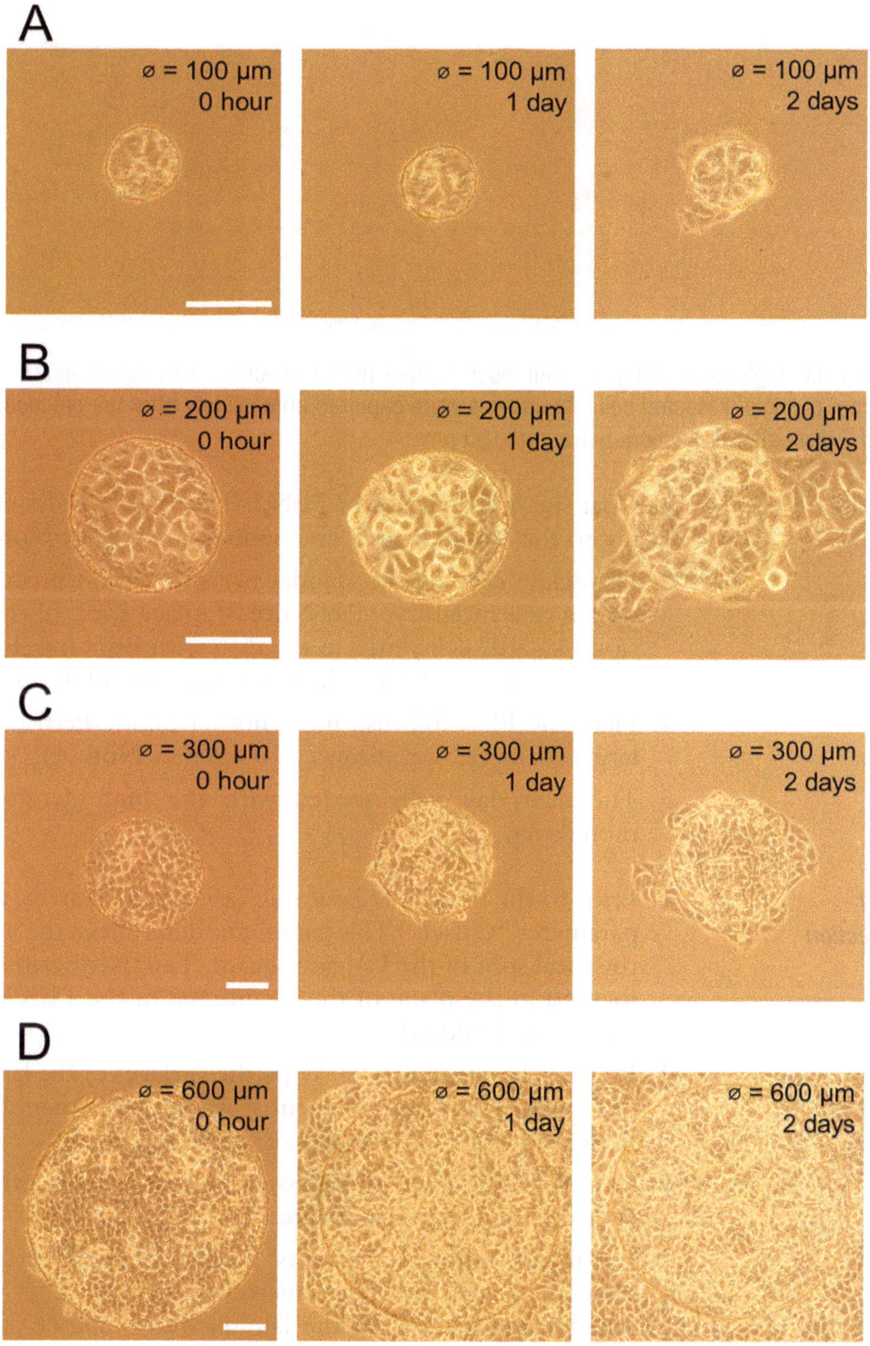

Fig. 3 Recultivation of cell monolayer fragments. This figure illustrates the formation of new colonies for samples with different diameters: 100 μm (**a**), 200 μm (**b**), 300 μm (**c**), and 600 μm (**d**). The images were captured at three timepoints after dissection: immediately after the procedure (0 hour), 1 day, and 2 days after dissection. Scale bars: 100 μm

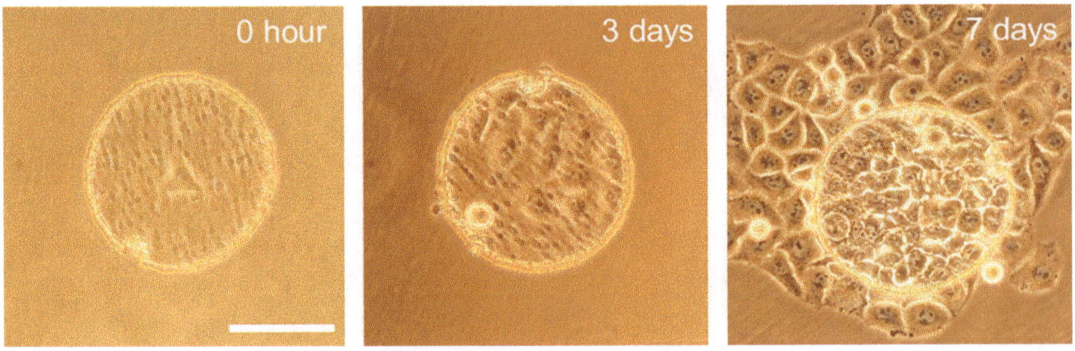

Fig. 4 Cloning of single cells using live cell laser capture microdissection. This figure shows new colony formation from a single dissected cell. The images were captured immediately after the procedure (0 hour), 3 days, and 7 days after dissection. Scale bar: 100 μm

2. Coat the inner surface of PEN foil with poly-D-lysine according to the manufacturer's recommendations (*see* **Note 2**).
3. Dissociate cells by the standard trypsinization procedure. To obtain cells monolayer (*see* **Note 3**), place 2 mL of cell suspension (4×10^5 cells/mL) in the dish with PEN foil. For single-cell capturing, use a less dense suspension (5000 cells/mL).
4. Place the PEN foil dish into another plastic Petri dish with a larger diameter, for instance, 90 mm (*see* **Note 4**).
5. The next day, cells seeded with the high density form a monolayer.

3.2 Laser Microdissection

1. Prior to the use of the laser microdissection system, adjust the parameter "Offset." This parameter determines the location of the focal spot of the UV laser beam. The laser beam should be focused in the plane of the membrane of the PEN dish filled with 1 mL of liquid.
2. Prepare the working area. Wipe the surfaces of the laser microdissection system and the incubator with 70% ethanol solution. Preferably use a UV to sanitize the room.
3. Switch on the laser microdissection instrument and start the laser microdissection software.
4. For recultivation experiments, sterilize the strip holder (supplemented with the laser microdissection system) with 70% ethanol and place it in the laminar flow cabinet. Insert two 8-well strips into the strip holder. Fill wells of the 8-well strips with 350 μL of culture medium (*see* **Note 5**) inside of the laminar flow cabinet. Leave one well of the strips empty and use this empty well during searching and cutting target areas containing cells (*see* **Note 6**). For downstream molecular profiling, insert 0.2 or 0.5 mL tubes into the tube collector (supplemented with the laser microdissection system). Fill lids

of the tubes with the extraction buffer suitable for subsequent downstream analysis. Similar to 8-well strips, leave one tube empty for the reasons mentioned above.

5. Place the strip holder with the two filled 8-well strips into a sterile container for transferring it to the laser microdissection instrument.
6. Remove medium from the PEN dish with cells. Removal and addition of media in the PEN dish should be carried out with caution. Avoid contact of micropipette tips with the PEN foil because of risk of its damage. To prevent cells in the PEN dish from drying out, keep them under a thin layer of liquid. Gently wash the cells with 2 mL of HBSS containing HEPES preheated to 37 °C. Remove the solution from the dish. Repeat the washing in 2 mL of fresh solution. Remove the solution from the dish and add 1 mL of fresh HBSS with HEPES (*see* **Note 7**). Keep the PEN dish inserted into the bigger plastic Petri dish to prevent contamination (*see* **Note 4**).
7. Bring the PEN dish inserted into a regular Petri dish and the strip holder in its sterile container to the laser microdissection instrument.
8. Use the respective functions of the laser microdissection software to load the microscope stage with the PEN dish and the substage with the collectors' holders.
9. Place the Petri dish without lid onto the stage of the laser microdissection instrument.
10. For recultivation experiments, place the strip holder with the two 8-well strips filled with culture medium into the substage of the laser microdissection instrument. For downstream molecular profiling, place the tube holder with tubes filled with the extraction buffer into the substage of the laser microdissection instrument.
11. Using the laser microdissection software, select the collector position corresponding to an empty well of the strip or an empty tube. The empty well or tube will be placed in the optical pathway of the microscope underneath the PEN dish.
12. Choose a circle region of interest on live images in the automatic mode "Draw and Cut" of the laser microdissection software. Use the laser settings presented in Table 1 to perform the dissection of circle areas with different diameters (*see* **Note 8**).
13. Press the "Start Cut" button. The focused laser beam will cut a circle region along the chosen contour (Fig. 1a, e).
14. The dissected sample is held by the force of the surface tension of the liquid (Fig. 1b, f). Using the option "Collector Device," switch the well or the tube underneath the PEN dish to the one with the medium or the extraction buffer.

Table 1
Optimal values of laser settings for live cell laser microdissection from HeLa cell monolayer determined for objectives with magnifications 10× and 20×. The parameter "Offset" should be determined empirically for an individual laser microdissection system. RU, in relative units [17]

Setting	Objective 10×	Objective 20×
Power (RU)	40	40
Aperture (RU)	1	1
Speed (RU)	6	6
Head current (%)	100	100
Pulse frequency (Hz)	120	120
Offset (RU)	–	–

Table 2
Optimal values of "Power" and "Aperture" settings for extracting of circle-shaped samples with different diameters using 10× and 20× objectives. RU, in relative units [17]

Objective	10×					20×				
Sample diameter (μm)	700	600	500	400	300	300	250	200	150	100
Power (RU)	60	50	45	40	40	40	40	40	50	45
Aperture (RU)	55	55	55	55	30	40	27	20	1	1

15. Use the manual mode "Move and Cut" of the laser microdissection software to shoot down the sample into the well or the tube. Apply the settings from Table 2 (*see* **Note 9**). Place the mouse pointer on the edge of the sample in the live image. Click the left mouse button to apply a shot laser pulse (Fig. 1c, g). The sample will fall into the well or the tube placed directly under the PEN dish (Fig. 1d, h, i). Repeat steps 16–20 to collect additional samples into separate wells or tubes. After completion of the laser microdissection procedure, appropriately dispose the PEN dish (*see* **Note 10**).

16. Remove the strip holder or the tube holder from the substage. For recultivation experiments, put the strip holder into the sterile container for transferring it back to the laminar flow cabinet. For molecular profiling, remove the tubes from the holder, carefully close them with the lids containing dissected samples, and centrifuge to collect liquid droplets with the dissected sample in bottoms of the tubes. Cell samples collected into tubes can be further proceeded for downstream molecular profiling according to respective protocols.

3.3 Cell Recultivation

1. Under sterile conditions, carefully remove the 8-well strips with excised cell samples from the strip holder and place them onto the stage of a stereomicroscope.
2. Using the reflected light mode of the stereomicroscope, identify the dissected cell samples floating on the surface of the liquid.
3. Using a micropipette, add a droplet of the medium on the surface of the floating samples to move them to the bottom of the well.
4. Place the strip into the strip plate, cover with the lid, and place in an incubator.

   Cultivate the cells using standard protocol (*see* **Note 5**).
5. Cell viability after laser microdissection may be assessed by double staining with propidium iodide (PI) and Sybr Green (SG) (Fig. 2) [14].
6. New colony formation from either cell monolayer fragments or single cells should be observed after laser microdissection, as shown in Figs. 3 and 4 (*see* **Note 11**).

4 Notes

1. The HeLa cell culture was used in the described procedure as one of the most widely used cell cultures. The protocol should be adapted for other cell cultures. For instance, a similar approach was used on CHO cells [15].
2. PEN foil pre-coating with poly-D-lysine increases adhesion of HeLa cells to the surface.
3. A previous study showed the quickest regrowth of colonies obtained from cell monolayers after laser dissection [16].
4. The risk of contamination during cell recultivation is increased due to the foil coming into contact with the outer dish space. To isolate the dish with cells, it is necessary to place the dish into a conventional sterile Petri dish.
5. The use of a conditioned medium is strongly recommended for the recultivation of samples with low densities of cells. This increases cell recultivation efficiency after laser microdissection.
6. Evaporating liquid from a well or a tube underneath of the PEN dish can interfere with the search for regions of interest and the microdissection process. Therefore, we recomend to leave one well empty and keep it under the PEN dish until the moment when gravity transfer is carried out. This empty well will also prevent foil and cell fragments from falling onto the condenser.

7. The height of the liquid layer above the cells is 580 ± 18 μm [17]. It is possible to cut and capture more than 20 samples during a 40-minute cutting session. If a longer time is required, completely remove the liquid and add 1 ml of fresh HBSS every 40 min. This procedure can be repeated at least two times for samples with different diameters without any loss of recultivation efficiency.
8. Correct laser settings are essential for cutting and transferring samples. The laser parameters used in this protocol have been empirically selected. The "Power," "Speed," and "Pulse frequency" parameters have been selected so that each pulse completely removes the foil over its entire thickness. Otherwise, the cavitation bubbles produced by the laser-induced plasma formation can ruin the procedure due to light scattering in the path of the laser beam. For "Aperture' parameter, the lower values are preferable as the laser focal spot will then have a large extension along the Z axis and the evaporation of the liquid during the dissection will not lead to a significant shift in focus.
9. For sample extraction in the manual "Move and Cut" mode, the minimal operating values of "Power" and the optimal "Aperture" were selected for different sample diameters at a "Pulse frequency" of 10 Hz. These values for "Power" and "Aperture" are presented in Table 2. The 800 μm diameter samples were not extracted at any settings, and for 700 μm, the efficiency was <20% (out of 100 attempts). The values presented in Table 2 are not suitable for complex forms of regions of interest.
10. PEN dishes from which cells have been cut should not be placed on the surface because they will spill due to holes left after laser microdissection.
11. No colonies form from 50 μm diameter samples. Approximately 50% of 100 μm samples give rise to growing colonies and 100% of larger diameter samples give rise to growing colonies [17].

Acknowledgments

This work was supported by the Russian Science Foundation, grant no. 23-75-30023.

References

1. Guo W, Hu Y, Qian J et al (2023) Laser capture microdissection for biomedical research: towards high-throughput, multi-omics, and single-cell resolution. J Genet Genomics 50: 641–651
2. Martin LBB, Nicolas P, Matas AJ et al (2016) Laser microdissection of tomato fruit cell and tissue types for transcriptome profiling. Nat Protoc 11:2376–2388
3. Clair G, Piehowski PD, Nicola T et al (2016) Spatially-resolved proteomics: rapid quantitative analysis of laser capture microdissected alveolar tissue samples. Sci Rep 6:39223
4. Rosenberger FA, Mädler SC, Thorhauge KH et al (2025) Deep Visual Proteomics maps proteotoxicity in a genetic liver disease. Nature:1–8
5. Ding J, Ahangari F, Espinoza CR et al (2019) Integrating multiomics longitudinal data to reconstruct networks underlying lung development. Am J Physiol-Lung Cell Mol Physiol 317:L556–L568
6. Dewez F, Martin-Lorenzo M, Herfs M et al (2019) Precise co-registration of mass spectrometry imaging, histology, and laser microdissection-based omics. Anal Bioanal Chem 411:5647–5653
7. Gautam V, Sarkar AK (2015) Laser assisted microdissection, an efficient technique to understand tissue specific gene expression patterns and functional genomics in plants. Mol Biotechnol 57:299–308
8. Niemz MH (2019) Laser-tissue Interactions: fundamentals and applications. Springer International Publishing, Cham
9. Burgemeister R (2005) New aspects of laser microdissection in research and routine. J Histochem Cytochem 53:409–412
10. Vogel A, Lorenz K, Horneffer V et al (2007) Mechanisms of laser-induced dissection and transport of histologic specimens. Biophys J 93:4481–4500
11. Podgorny OV, Polina NF, Babenko VV et al (2015) Isolation of single *Chlamydia*-infected cells using laser microdissection. J Microbiol Methods 109:123–128
12. Kim Y, Jang H, Seo K et al (2022) Cell position within human pluripotent stem cell colonies determines apical specialization via an actin cytoskeleton-based mechanism. Stem Cell Rep 17:68–81
13. Malolina EA, Kulibin AY (2019) The rete testis harbors Sertoli-like cells capable of expressing DMRT1. Reproduction 158:399–413
14. Garner DL, Johnson LA (1995) Viability Assessment of mammalian sperm using SYBR-14 and propidium Iodide1. Biol Reprod 53: 276–284
15. Horneffer V, Linz N, Vogel A (2007) Principles of laser-induced separation and transport of living cells. J Biomed Opt 12:054016
16. Stich M, Thalhammer S, Burgemeister R et al (2003) Live Cell Catapulting and Recultivation. Pathol - Res Pract 199:405–409
17. Podgorny OV (2013) Live cell isolation by laser microdissection with gravity transfer. J Biomed Opt 18:1

Chapter 3

From Laser Microdissection to Spatial Glycomics: Lectin Microarray Protocols for Tissue Glycome Mapping with High-End Scanners

Patcharaporn Boottanun, Sayaka Fuseya, Atsushi Kuno, and Chiaki Nagai-Okatani

Abstract

Formalin-fixed paraffin-embedded (FFPE) tissues offer key advantages such as broad availability, long-term preservation, and retention of tissue architecture, providing the substrate on which laser microdissection (LMD) turns spatial glycomics into practice. Integrating LMD into the lectin microarray workflow (LMD-LMA) enables precise and reproducible glycomic profiling of microscale tissue regions (~0.1 mm^2 areas, 5 μm thickness), thereby revealing intra- and inter-tissue-specific glycosylation patterns.

Here, we present detailed protocols for LMD-LMA methods optimized with a high-end evanescent-field fluorescence scanner (GSR2300), which improves sensitivity by detecting weak signals while reducing background noise. The workflow can be implemented through three approaches based on the dissection strategy: Method 1, morphology-guided dissection; Method 2, probe-guided dissection on serial sections; and Method 3, direct dissection from probe-stained sections. Regardless of the dissection method, tissue fragments follow a common LMA pipeline comprising protein extraction, Cy3 labeling, lectin–glycoprotein interaction on the array, fluorescence scanning, and data qualification. The resulting "tissue glycome mapping" datasets are shareable via LM-GlycoRepo and visualized through LM-GlycomeAtlas under FAIR (Findable, Accessible, Interoperable, Reusable) principles. Crucially, this protocol is plug-and-play with existing LMD pipelines: Microdissected regions can be mirrored across serial sections for spatial transcriptomics or proteomics, enabling co-registered datasets in which glycan motifs serve as anchors for biological interpretation. We encourage the broader spatial omics community to enrich current workflows with glycan information at minimal overhead.

Key words Glycoproteins, Glycomics, Lectin microarray, Spatial glycomics, Tissue glycome mapping

1 Introduction

For researchers already using laser microdissection (LMD) on formalin-fixed paraffin-embedded (FFPE) tissues, spatial glycomics can be implemented as a drop-in extension of existing workflows. Among biological samples, FFPE tissues offer several advantages mainly in terms of availability, long-term storage, and visualization

Roberta Noberini (ed.), *Laser Capture Microdissection: Methods and Protocols*, Methods in Molecular Biology, vol. 3015,
https://doi.org/10.1007/978-1-0716-5154-4_3,

of tissue architecture. Consequently, FFPE is widely used for the current trend of spatial omics analyses [1], such as transcriptomics [2], proteomics [3], and glycomics [4], because it provides spatially informative data, including pathological morphology and cellular communication [5]. LMD has been introduced for high-resolution omics analysis to enable accurate and precise collection of tissue fragments from microscale regions [6, 7].

Glycosylation is a posttranslational modification in which sugar chains (glycans) are added to proteins, generating glycoproteins with diverse biological functions [8]. Aberrant glycosylation has been reported in many diseases, making the identification of glycan structures important for biomarker discovery and therapeutic targeting [9–11]. Because glycans are synthesized through diverse, context-dependent enzymatic processes, they exhibit structural diversity and complexity and cannot be directly inferred from the genome [12]. Consequently, comprehensive characterization must rely on direct analytical measurements using advanced technologies capable of resolving glycan complexity. Aberrant glycosylation is a hallmark of pathogenesis, yet methods for spatial glycomics—guided by tissue morphology or molecular markers—have lagged behind transcriptomics and proteomics [13]. Therefore, lectin microarray (LMA) provides a practical platform for glycomic profiling that can complement existing omics approaches and enable morphology- or probe-guided spatial glycomics when combined with LMD [13].

LMA is widely used for glycomic profiling based on the defined binding specificities of lectins [14]. In LMA, multiple lectins are immobilized on solid surfaces such as glass slides, enabling high-throughput analysis of both *N*- and *O*-glycans simultaneously. Mass spectrometry (MS)-based approaches are the gold standard for glycan analysis, providing detailed structural composition but requiring enzymatic cleavage and often lacking sensitivity for low-abundance glycans. In contrast, LMA offers higher sensitivity without glycan release but yields motif-level profiles rather than full structural elucidation, making it complementary to MS [13]. This versatility allows LMA to be applied to a wide range of biological samples, including serum, crude, or purified glycoproteins from whole cells or tissues, and even FFPE tissue sections [15].

For spatial glycomic analysis using FFPE tissue sections, a procedure using LMD-assisted LMA (LMD-LMA) has been standardized to ensure reliability and reproducibility of the obtained data [16]. This optimized workflow can be implemented through three distinct approaches based on the tissue dissection strategy prior to LMA (Fig. 1a) [17]. In Method 1, hematoxylin-stained FFPE tissue sections on polyethylene naphthalate (PEN) membrane slides are directly dissected, enabling broad morphology-based glycomic analysis; in Method 2, regions are identified by specific molecular probes (e.g., antibodies or lectins) on one section

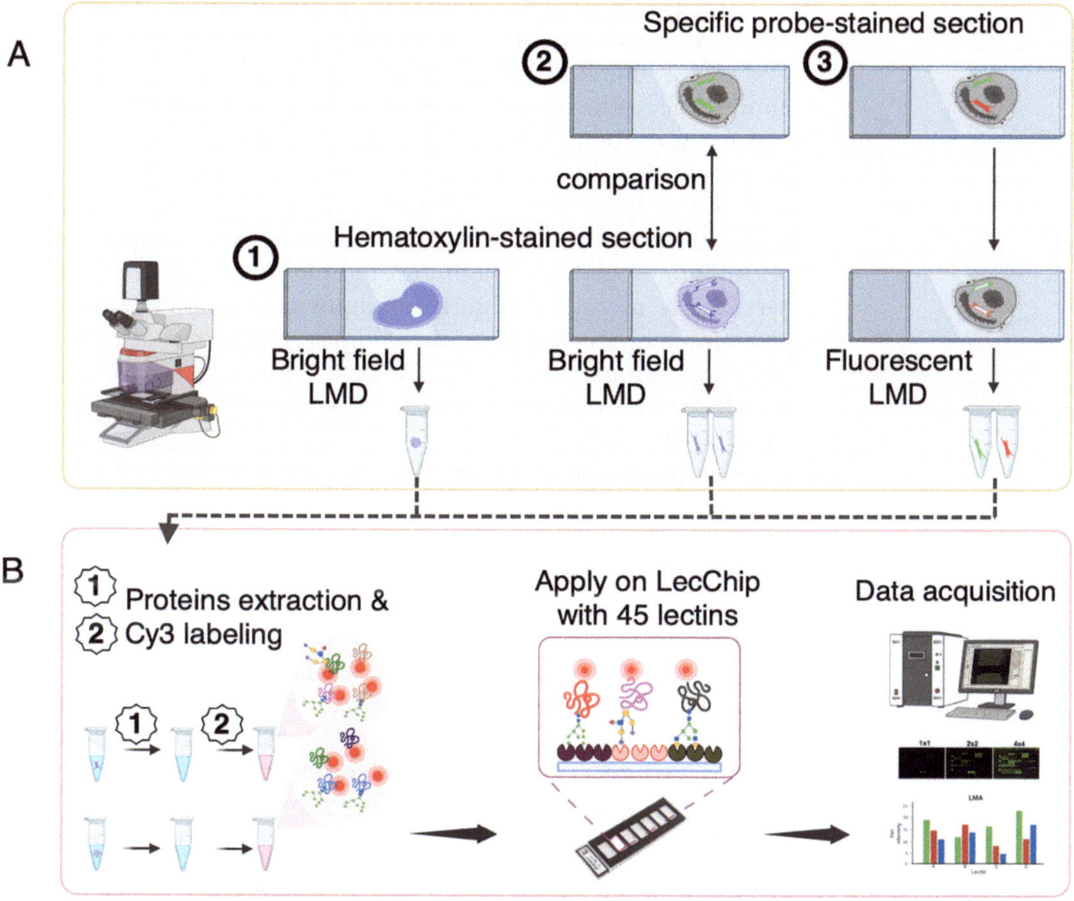

Fig. 1 Schematic workflow of laser microdissection-assisted lectin microarray (LMD-LMA). (**a**) Three tissue dissection methods prior to LMA are shown. In Method 1, hematoxylin-stained tissue sections on PEN membrane slides are directly subjected to LMD. In Method 2, specific molecular probes are used to identify target regions on one section, while dissection is performed on a hematoxylin-stained serial section to avoid probe interference with analytes. In Method 3, targeted dissection is performed directly on probe-stained tissue sections mounted on PPS-membrane slides, enabling precise sampling down to the cell-type level. (**b**) In all three methods, collected tissue fragments undergo the same LMA workflow: protein extraction, Cy3 labeling, interaction between glycosylated proteins and immobilized lectins on the microarray chip, and scanning with an LMA scanner to obtain glycomic profiles

while dissection is performed on a hematoxylin-stained serial section to avoid probe interference; and in Method 3, probe-stained sections mounted on polyphenylene sulfide (PPS) are dissected directly, allowing high-precision sampling down to the cellular level [17]. Regardless of the approach, all workflows share the same downstream pipeline: protein extraction, Cy3 labeling, application to the lectin microarray where glycosylated proteins interacted with lectins, and scanning to obtain glycomic profiles (Fig. 1b).

Because most LMAs employ fluorescence-based detection, microarray scanners are essential components of the workflow. The GSR2300, as an advanced scanner using the evanescent-field fluorescence principle and digital binning, is optimized for high sensitivity, enabling reliable detection of weak signals in small tissue fragments (0.1 mm^2, 5 μm thickness) and supporting comparative profiling of inter- and intra-tissue samples [15]. In contrast, the GSL2200 provides rapid one-shot scanning in a compact design, prioritizing speed and high throughput over sensitivity, and is particularly useful for high-throughput and time-efficient spatial glycomic analysis. Together, these high-end scanners offer complementary options: The GSR2300 excels in sensitivity for small or weakly stained regions, whereas the GSL2200 emphasizes speed and usability.

The feasibility of the LMD-LMA method has been demonstrated in "tissue glycome mapping" to reveal region-specific patterns of protein glycosylation both intra- and inter-tissues in normal mice [18]. This approach also differentiates glycan structures specifically expressed in the fibrotic cardiac tissues of dilated cardiomyopathy model mice [9]. Additionally, visualized histological images of more than 500 tissue glycomic profiles across multiple tissues are publicly available on the LM-GlycomeAtlas [18, 19] (https://glycosmos.org/lmglycomeatlas). Recently, LM-GlycoRepo (https://lm-glycorepo.glycosmos.org/lm_glycorepo/) is launched as a repository for the deposition of tissue glycome mapping datasets to support multimodal data integration [20]. In keeping with the open science movement, the use of Findable, Accessible, Interoperable, and Reusable (FAIR) repositories is essential to ensure the accessibility and reproducibility of omics data. Platforms such as LM-GlycoRepo and LM-GlycomeAtlas enable the standardized deposition and visualization of LMA-based tissue glycomics datasets, facilitating cross-omics comparison and integration with other omics resources.

In this chapter, we present a step-by-step protocol for LMD-LMA, covering the entire workflow, from the preparation and staining of FFPE tissue sections to laser microdissection, protein extraction, fluorescence labeling, and lectin microarray analysis using GSR2300. The Materials section details the reagents, equipment, and preparation procedures required for each step. The Methods section provides practical guidance for implementing the three aforementioned LMD-LMA workflows, including data acquisition and processing. As the detailed procedure for Method 3 has been described in our previous methodological publication [21], this article focuses on providing in-depth protocols for Methods 1 and 2. This protocol was designed for both inter- and intra-tissue glycomic profiling and can be applied to various archived FFPE specimens.

2 Material

All solutions are prepared using ultrapure water and analytical grade reagents. All reagents and materials are stored at room temperature unless otherwise specified. For the commercially available kits, follow the manufacture's protocol. Ensure that sufficient reagents are prepared for the complete analysis to minimize inter-batch variation. Follow all safety and waste disposal regulations.

2.1 Preparation of FFPE Tissue Sections for Tissue Dissection

1. FFPE tissue blocks (*see* **Note 1**).
2. Glass slides; MAS-coated glass slide for reference-specific probe staining and polyethylene naphthalate (PEN) membrane glass slides for preparation of tissue sections for staining with hematoxylin.
3. 0.1% (w/v) poly-L-lysine solution.
4. Microtome equipped with disposable blades.

2.2 Hematoxylin Staining of FFPE Tissue Sections-PEN Membrane Slide

1. Deparaffinization and rehydration solvents: xylene and 100%, 90%, 70%, and 50% (v/v) ethanol.
2. Mayer's hematoxylin solution.
3. Fluorescence microscope, for bright-field and fluorescent staining observation.

2.3 Fluorescence Histochemical Staining of FFPE Tissue Sections

1. Solvents for deparaffinization and rehydration (*see* Subheading 2.2, Material 1).
2. Glass slides (*see* Subheading 2.2, Material 2).
3. Antigen retrieval buffer: Target Retrieval Solution, citrate (pH 6.0) (*see* **Note 2**).
4. 10 mM phosphate-buffered saline (PBS).
5. A streptavidin-biotin blocking kit containing streptavidin and a biotin solution was used.
6. Carbo-free blocking solution (Vector Laboratories, Burlingame, CA, USA). The working solution is obtained by diluting the concentrated stock solution ten-fold with water.
7. Heat-resistant plastic Coplin jar.
8. PAP pen (hydrophobic barrier pen).
9. Biotinylated lectins or antibodies.
10. Fluorescein isothiocyanate (FITC)-labeled lectin of interest, which is used for fluorescence double-staining analysis (*see* **Note 3**).
11. Alexa Fluor 647-conjugated streptavidin, which is used for fluorescence double staining using a lectin and an antibody (*see* **Note 3**).

12. Antifade mountant.
13. Fluorescence microscope (*see* Subheading 2.2, Material 3).

2.4 LMD-Assisted Tissue Dissection

1. LMD system (LMD7000) attached to a microscope (DM6000B, Leica Microsystems, Wetzlar, Germany).
2. 0.5 mL tissue collection tube.

2.5 Tissue Protein Extraction for LMA Analysis

1. 10 mM citrate buffer (pH 6.0): Prepare a 0.1 M citric acid solution by dissolving 2.1 g of citric acid monohydrate in 100 mL of water. Prepare a 0.1 M sodium citrate solution by dissolving 14.7 g of trisodium citrate dihydrate in 100 mL of water. Mix 3.6 μL of 0.1 M citric acid solution, 16.4 μL of 0.1 M sodium citrate solution, and 180 of μL water (*see* **Note 4**) to prepare 200 μL of working buffer per sample. The pH of the final buffer was 6.0 ± 0.2. Use the prepared buffer within one day.
2. Dulbecco's PBS tablet (D-PBS).
3. 0.5% (v/v) Nonidet P-40 (NP-40) in D-PBS: Mix 5 μL NP-40 and 995 μL D-PBS to obtain 1000 μL of the working solution (*see* **Note 5**). Use the prepared solution within 1 day.
4. Ultrasonic bath sonicator (*see* **Note 6**).
5. Low-retention tips and protein low bind tubes (*see* **Note 7**).

2.6 Fluorescence Labeling and LMA Analysis of Tissue Protein Samples

1. Preparation of working Cy3-succinimidyl ester (Cy3-SE) dye (10 μg protein equivalent of Cy3-SE/tube): Dissolve 1 mg protein equivalent of Cy3-SE (GE Healthcare, Buckinghamshire, UK) in N,N-dimethylformamide which has been dehydrated using molecular sieves prior to use. Aliquot 5 μL of the resulting dye solution into each tube of an eight-strip 0.2 mL tube set using a 1–10 μL autopipette and dry the aliquots completely using a cold trap-vacuum centrifuge evaporator. Store the working dye with a dehydrating agent at 4 °C in the dark (*see* **Note 8**).
2. Probing buffer: 25 mM Tris–HCl, pH 7.5, containing 137 mM NaCl, 2.7 mM KCl, 500 mM glycine, 1 mM $CaCl_2$, 1 mM $MnCl_2$, and 1% Triton X-100. Store at 4 °C.
3. Lectin array chips (LecChip ver.1.0; GlycoTechnica, Yokohama, Japan) comprising triplicate spots of 45 lectins (Fig. 2 and Table 1) (*see* **Note 9**).
4. Humidified shading chamber.
5. Low-retention tips and tubes (*see* Subheading 2.5, **Item 5**).
6. Incubator, with the temperature set to 20 °C.
7. Rotator shaker placed inside the incubator (*see* Subheading 2.6, **Item 6**).

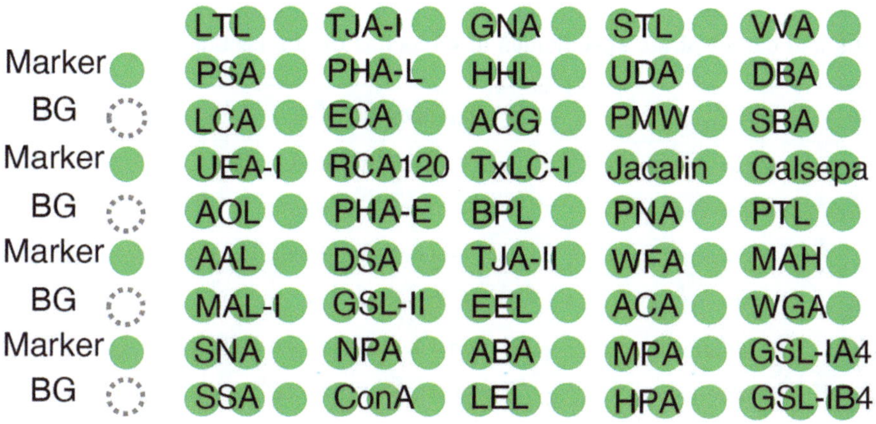

Fig. 2 Illustration of the lectin microarray (LecChip ver 1.0) showing the 45-lectin panel on one well of the chip. Abbreviations of the lectins and their specificities are listed in Table 1. BG indicates the background areas used for signal normalization and analysis

8. Evanescent-field fluorescence scanner (GlycoStation Reader 2300; Emukk, Yokohama, Japan).
9. LMA data analysis software (GlycoStation Tools Pro Suite 1.5 GlycoTechnica, Kuwana, Japan).

3 Methods

3.1 Preparation of Hematoxylin-Stained Tissue Sections

This pretreatment method is used to examine the tissue morphology under a light microscope for LMD-assisted tissue collection. For example, this protocol has been applied to inter-tissue glycomic profiling using FFPE kidney sections from normal mice. The significant lectins identified using this method can be used as specific probes in Methods 2 and 3.

3.1.1 Preparation of Tissue Sections

1. Apply 80 μL of 0.1% poly-L-lysine solution to a PEN membrane glass slide and allow it to air-dry for 12–24 h (*see* **Note 10**).
2. Slice the FFPE tissue block using a microtome. Float the resulting 5 μm thick sections on the surface of water preheated at 42 °C in a water bath to flatten and mount on a poly-L-lysine-coated slide.
3. Dry the slide overnight on a hot plate at 42 °C.

3.1.2 Hematoxylin Staining

1. Deparaffinize and rehydrate the slide by stepwise immersion in the following solutions for the specified times: xylene, three times for 5 min each; 100% ethanol, three times for 5 min each; 90% ethanol, once for 3 min; 70% ethanol, once for 3 min; 50% ethanol, once for 3 min; and water, twice for 1 min each.

Table 1
Abbreviations and carbohydrate specificities of 45 lectins on the LecChip

No.	Lectin	Origin	Binding specificity[a]
1	LTL	*Lotus tetragonolobus*	Fucα1–3GlcNAc, Sia-Lex, and Lex
2	PSA	*Pisum sativum*	Fucα1-6GlcNAc and α-man
3	LCA	*Lens culinaris*	Fucα1-6GlcNAc, α-man, and α-Glc
4	UEA-I	*Ulex europaeus*	Fucα1–2LacNAc
5	AOL	*Aspergillus oryzae*	Terminal α-Fuc, Sia-Lex, and Lex
6	AAL	*Aleuria aurantia*	Terminal α-Fuc, Sia-Lex, and Lex
7	MAL-I	*Maackia amurensis*	Siaα2-3Gal
8	SNA	*Sambucus nigra*	Siaα2-6Gal/GalNAc
9	SSA	*Sambucus sieboldiana*	Siaα2-6Gal/GalNAc
10	TJA-I	*Trichosanthes japonica*	Siaα2-3Galβ1-4GlcNAcβ-R
11	PHA-L	*Phaseolus vulgaris*	Tri- and tetra-antennary complex-type N-glycan
12	ECA	*Erythrina crista-galli*	Lac/LacNAc
13	RCA120	*Ricinus communis*	Lac/LacNAc
14	PHA-E	*Phaseolus vulgaris*	NA2 and bisecting GlcNAc
15	DSA	*Datura stramonium*	(GlcNAc)n, polyLacNAc, and LacNAc (NA3, NA4)
16	GSL-II	*Griffonia simplicifolia*	Agalactosylated N-glycan
17	NPA	*Narcissus pseudonarcissus*	Non-substituted α1-6Man
18	ConA	*Canavalia ensiformis*	α-Man (inhibited by presence of bisecting GlcNAc)
19	GNA	*Galanthus nivalis*	Non-substituted α1-6Man
20	HHL	*Hippeastrum hybrid*	Non-substituted α1-6Man
21	ACG	*Agrocybe cylindracea*	Siaα2-3Galβ1-4GlcNAc
22	TxLC-I	*Tulipa gesneriana*	Man3, bi- and tri-antennary complex-type N-glycan, and GalNAc
23	BPL	*Bauhinia purpurea alba*	Galβ1–3GalNAc and NA3, NA4
24	TJA-II	*Trichosanthes japonica*	Fucα1–2Gal, β-GalNAc > NA3, NA4
25	EEL	*Euonymus europaeus*	Galα1–3[Fucα1–2]Gal > Galα1–3Gal
26	ABA	*Agaricus bisporus*	Galβ1–3GalNAcα-Thr/Ser (T) and sialyl-T
27	LEL	*Lycopersicon esculentum*	(GlcNAc)n and polyLacNAc
28	STL	*Solanum tuberosum*	(GlcNAc)n and polyLacNAc
29	UDA	*Urtica dioica*	(GlcNAc)n and polyLacNAc
30	PWM	*Phytolacca americana*	(GlcNAc)n and polyLacNAc
31	Jacalin	*Artocarpus integrifolia*	Galβ1–3GalNAcα-Thr/Ser (T) and GalNAcα-Thr/Ser (Tn)

(continued)

Table 1
(continued)

No.	Lectin	Origin	Binding specificity[a]
32	PNA	*Arachis hypogaea*	Galβ1–3GalNAcα-Thr/Ser (T)
33	WFA	*Wisteria floribunda*	Terminal GalNAc (e.g., GalNAc β1-4GlcNAc)
34	ACA	*Amaranthus caudatus*	Galβ1–3GalNAcα-Thr/Ser (T)
35	MPA	*Maclura pomifera*	Galβ1–3GalNAcα-Thr/Ser (T) and GalNAcα-Thr/Ser (Tn)
36	HPA	*Helix pomatia*	Terminal GalNAc
37	VVA	*Vicia villosa*	α-, β-linked terminal GalNAc and GalNAcα-Thr/Ser (Tn)
38	DBA	*Dolichos biflorus*	GalNAcα-Thr/Ser (Tn) and GalNAcα1–3GalNAc
39	SBA	*Glycine max*	Terminal GalNAc (especially GalNAcα1–3Gal)
40	Calsepa	*Calystegia sepium*	Man and Maltose
41	PTL-I	*Psophocarpus tetragonolobus*	α-GalNAc and Gal
42	MAH	*Maackia amurensis*	Siaα2-3Galβ1–3[Siaα2-6GalNAc]α-R
43	WGA	*Triticum vulgaris*	(GlcNAc)n and multivalent Sia
44	GSL-I-A4	*Griffonia simplicifolia*	α-GalNAc and GalNAcα-Thr/Ser (Tn)
45	GSL-I-B4	*Griffonia simplicifolia*	α-Gal

[a]Binding specificities are based on Lectin Frontier Database (LfDB; https://acgg.asia/lfdb2)

2. Immerse the slide in Mayer's hematoxylin solution for 1–10 min and wash under running tap water for 10 min (*see* **Note 11**).
3. Dry the slide by gently blowing air (e.g., using a hair dryer) (*see* **Note 12**). Store in the dark until use (*see* **Note 13**).
4. Scan the high-resolution image of the entire section under bright-field optics using a BZ-710 microscope, before being subjected to LMD. The image will be used for the annotation of dissected areas (*see* **Note 14**).

3.2 Preparation of Specific Probe Fluorescent-Stained Tissue Section (for Method 2)

This fluorescently stained section will be used as a reference for targeted glycomic profiling (Method 2), using a lectin or antibody against a specific molecular marker as the staining probe (*see* **Note 15**) [17]. As an example, WFA was used to detect aberrant fibrosis-related glycosylation in heart tissue samples from both control and diseased rats, along with WGA as a membrane marker. LMD-assisted tissue collection will be performed on a serial set of hematoxylin-stained sections (*see* Subheadings 3.1 and 3.3). In both Methods 1 and 2, the subsequent protocols following pre-treatment overlap, as described below (*see* Subheadings 3.4, 3.5, and 3.6).

3.2.1 Preparation of Tissue Sections

1. Slice an FFPE tissue block using the sliding microtome and then mount on an MAS-coated glass slide (*see* Subheading 3.1.1, **Step 2**).
2. Dry the slide overnight on a hot plate at 42 °C.

3.2.2 Specific Probe Fluorescent Staining

1. *Deparaffinization and antigen retrieval*: Deparaffinize and rehydrate sections as described in Sect. 3.1.2 (**steps 1**). For antigen retrieval, place slides in a heat-resistant plastic Coplin jar containing citrate buffer (pH 6.0) and autoclave for 10 min at 110 °C. Cool the slides to room temperature in buffer (*see* **Note 16**) and rinse three times with PBS. Remove residual water with Kimwipes and outline the tissue with a PAP pen.
2. *Blocking*: To block endogenous biotin, incubate sections with streptavidin solution for 15 min and wash three times with PBS. Then incubate the samples in a biotin solution for 15 min and wash again (*see* **Note 17**). Block nonspecific binding with a working carbo-free blocking solution for 1 h, followed by three washes with PBS.
3. *Staining with target probes*: Prepare biotinylated WFA (2 μg/mL in PBS) and FITC-conjugated WGA (2 μg/mL in PBS) solutions and incubate sections with this solution overnight at 4 °C in the dark. Wash three times with PBS.
4. *Secondary staining*: Incubate with Alexa Fluor 647-conjugated streptavidin (2 μg/mL in PBS) for 1 h in the dark, then wash three times with PBS.
5. *Mounting and imaging*: Bring ProLong Diamond antifade reagent to room temperature. Remove excess buffer, apply 20–60 μL mountant to a coverslip, and gently place over the section, avoiding air bubbles. Allow the mountant to set at room temperature for 1 h or 4 °C overnight in the dark. Examine using a fluorescence microscope with appropriate filters.

3.3 LMD-Assisted Tissue Dissection

1. Attach the hematoxylin-stained section and tissue collection tubes to the holders of the LMD system (*see* **Note 18**).
2. Collect tissues using LMD as follows:

   For Method 1: Dissect tissue fragments of the glomerulus and tubular (total area of 0.1 mm^2 per tube for each region) in the kidney under bright-field optics (*see* **Note 19**) using the following conditions: magnification, 6.3–20; laser power, 15–30; aperture, 6; and speed, 1 (*see* **Note 20**). Collect the tissue fragments into the caps of collection tubes.

For Method 2: Prior dissection aligns the hematoxylin-stained sections with WFA-stained sections for targeted areas (WFA^+ and WFA^-) annotation (*see* **Note 21**). After defining targeted regions, dissect tissue fragments of WFA^+ and WFA^- areas (total area of 0.5 mm^2 per tube) under bright-field optics and collect tissue fragments using the same conditions as above.

3. Perform the same procedure for one tube containing the PEN membrane fragment from the area without tissue as a negative control to evaluate potential contamination derived from the PEN membrane.
4. Close the cap gently while positioned upside down (with the cap facing downward).
5. After completing all the dissection procedures, capture a high-resolution image of the entire section as described above. This image will be used to annotate the dissected areas (*see* Subheading 3.1.2, **Step 4**).
6. Figure 3 shows sample images of tissue dissection from hematoxylin-stained sections using Method 1. Figure 4 shows sample images of tissue dissection from hematoxylin-stained sections of Method 2, with the overlaid image after dissection of the WFA-stained section.

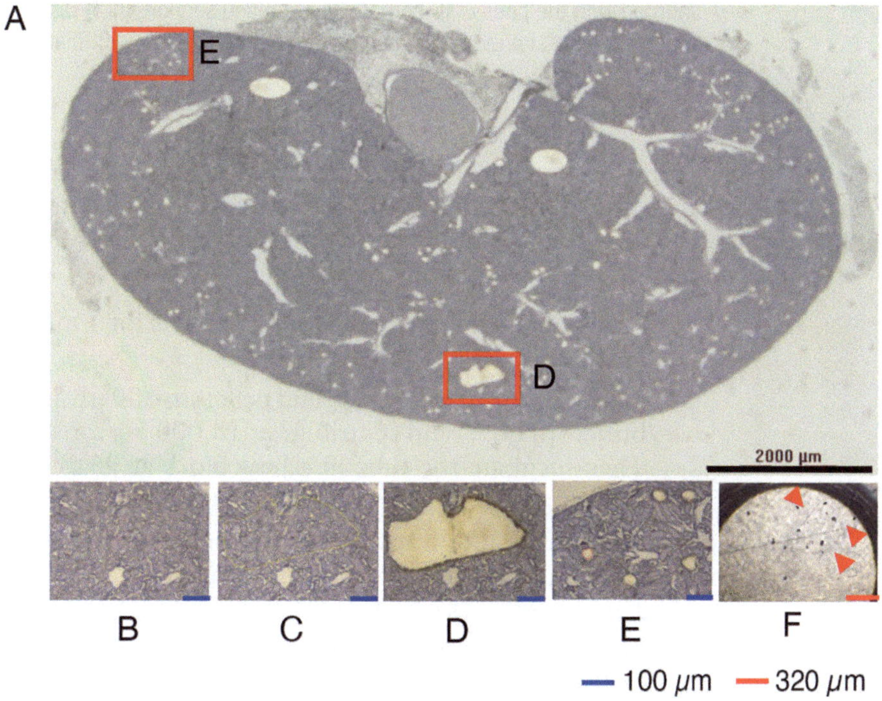

Fig. 3 Demonstration of LMD-assisted tissue dissection of a section stained with hematoxylin (Method 1). Panels A–F indicate (**a**) whole-slide image after LMD, (**b**) area of interest, (**c**) drawing of the dissection area before cutting, (**d**) tubular area after dissection, as shown in the whole-slide image, (**e**) glomerulus after dissection, and (**f**) confirmation of dissected fragments under bright-field microscopy

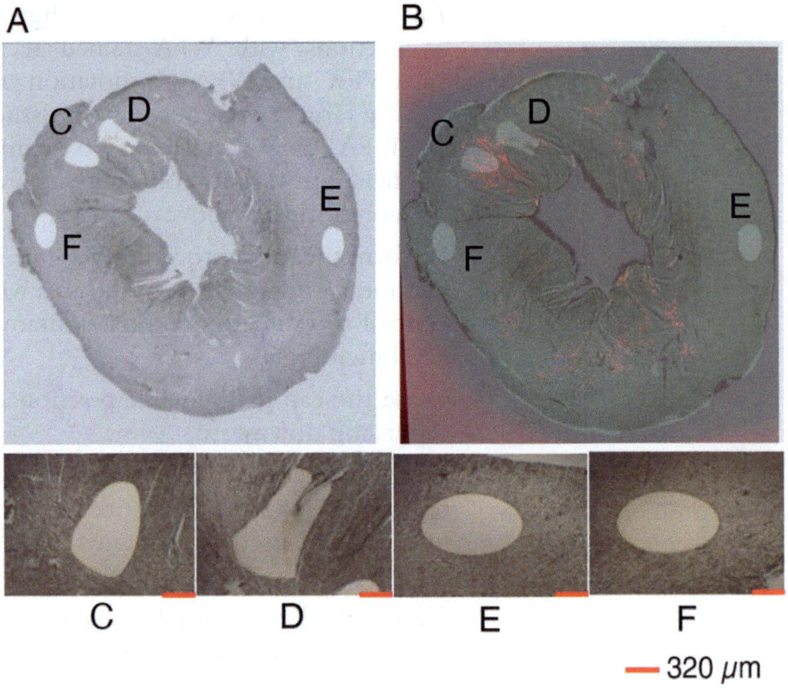

Fig. 4 Example of targeted tissue dissection using a guided, specific probe-stained section (Method 2). (**a**) The whole-slide image of a hematoxylin-stained section after LMD. (**b**) The whole-slide image of a specific probe-stained section merged with the post-LMD hematoxylin-stained section. (**c–f**) Dissected areas of a hematoxylin-stained section corresponding to regions identified on the whole-slides images

3.4 Protein Extraction and Fluorescent Labeling

Consistent conditions and maximized efficiency in protein extraction and labeling directly affect the accuracy of glycomic profiles. It is important to conduct these procedures simultaneously to ensure a reliable comparative analysis.

1. Centrifuge tissue fragments at 18,000 × *g* for 1 min at 4 °C (*see* **Note 22**).
2. *Heat denaturation*: Resuspend pellets in 200 μL 10 mM citrate buffer (pH 6.0) and centrifuge at 18,000 × *g* for 1 min at 4 °C. Then incubate the tube in a heat block at 95 °C for 1 h and centrifuge at 18,000 × *g* for 1 min at 4 °C (*see* **Note 23**).
3. *Wash*: Carefully remove 190 μL supernatant, add an equal volume of D-PBS, vortex, and centrifuge at 18,000 × *g* for 1 min at 4 °C (*see* **Note 24**).
4. *Detergent treatment*: Remove 190 μL supernatant, add 10 μL D-PBS containing 1% NP-40, vortex, and chill on ice for 1 min (*see* **Note 25**).

5. *Protein extraction*: Hold the tube by hand and sonicate 3 s at 20 °C and chill on ice for 1 min; repeat twice (three cycles total). Incubate on ice for 1 h (*see* **Note 26**). Centrifuge at 18,000 × *g* for 1 min at 4 °C, collect 10 μL of supernatant into a low-retention tube (*see* **Note 27**), and adjust to 20 μL with D-PBS. Store at −30 °C if not used immediately (*see* **Note 28**).
6. *Labeling*: Mix 20 μL protein extract with Cy3-SE dye (10 μg protein equivalent) by pipetting and incubate for 1 h at room temperature in the dark (*see* **Note 30**).
7. *Quenching*: Add 80 μL probing buffer and incubate for 2 h in the dark. Store Cy3-labeled protein sample on ice for same-day analysis or at −30 °C until use (*see* **Note 29**).

3.5 LMA Analysis of Cy3-Labeled Protein Samples

1. *Prewash of chip*: Bring the lectin array chip (*see* **Note 30**) to room temperature and remove it from the storage bag. Wash each well three times with 60 μL of probing buffer (*see* **Note 31**). Add 60 μL of the probing buffer to each well and place the chip in a humidified light-shaded chamber until use.
2. *Sample application and incubation*: Completely discard the probing buffer in the chip and immediately apply the samples to each well (60 μL/well) (*see* **Note 32**). Incubate the chip in the humidified light-shaded chamber overnight at 20 °C with shaking at 60 rpm using a rotator shaker.
3. *Washing*: Discard the sample mixture and wash the chip three times with 60 μL/well probing buffer.
4. *Scanning procedure*: Add 60 μL/well of the probing buffer to each well after the final wash (*see* **Note 33**), and scan the chip using the evanescent-field fluorescence scanner under the following conditions: exposure times of 0.5, 1, 2, 3, 4, 5, and 7 s with a digital binning mode of 1 × 1; 0.5, 1, 2, and 3 s for 2 × 2; and 0.1, 0.25, 0.5, and 1 for 4 × 4 (*see* **Note 34**). This serial scanning results in obtaining seven images for the 1 × 1 binning mode and four images for 2 × 2 and 4 × 4 per chip.
5. *Data illustration*: Figure 5a shows an example scanning image of the lectin array chip and the proportion ratio of net intensity in 1 × 1, 2 × 2, and 4 × 4 binning mode at 0.5 sec, along with a similar glycomic profile after normalization (Fig. 5b).

3.6 LMA Data Processing

1. Calculate the net intensity of each spot on the lectin array chip, using the GlycoStation Tool Pro software (*see* **Note 35**), with the grid size set at 17. Apply the same grid to all exposure times and binning modes for a given chip.
2. Export text-format report files containing net intensity values for each exposure condition for each sample.

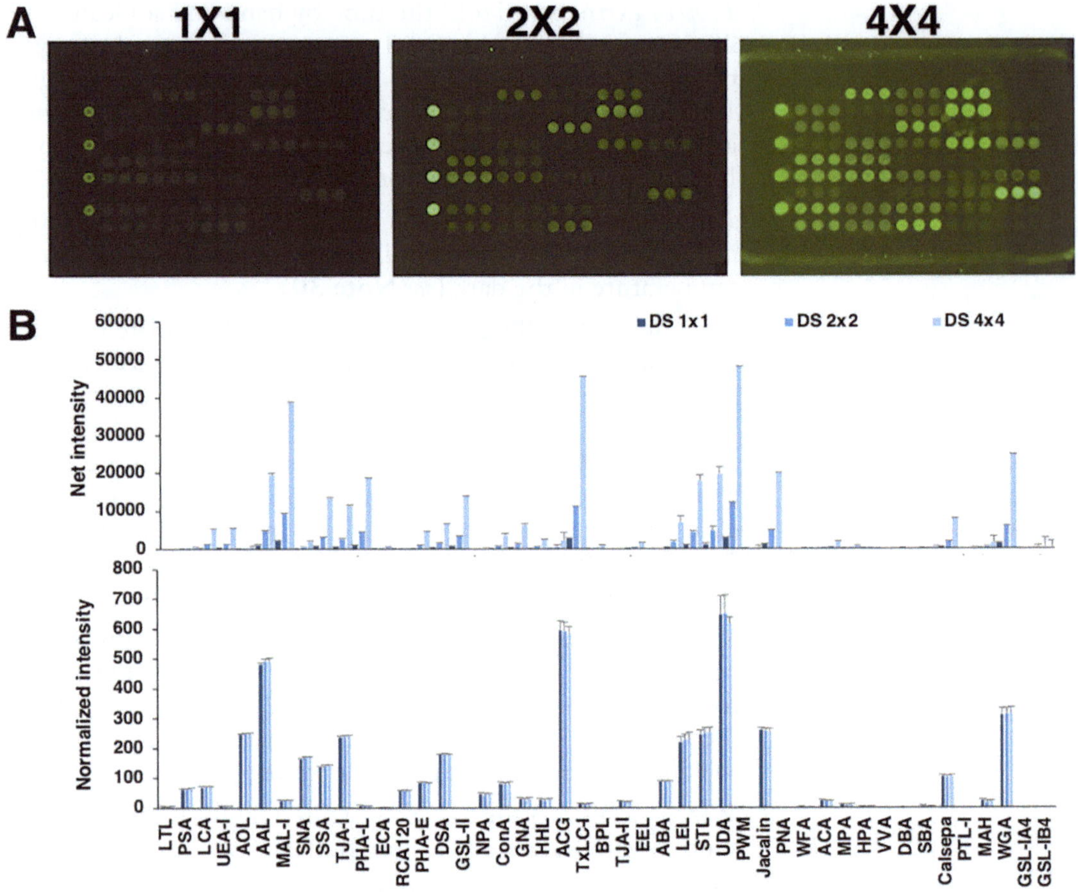

Fig. 5 Data acquisition from lectin microarray analysis. (**a**) LMA-scanned images at 0.5 sec exposure showing signal intensities for 1 × 1, 2 × 2, and 4 × 4 binning mode, with signal intensity increasing as binning mode increases. (**b**) The net intensity of one sample, presented as proportion ratios for 1 × 1, 2 × 2, and 4 × 4 binning modes, along with the corresponding normalized glycomic profiles demonstrating similarity across binning modes

3. Compile the net intensity data from the scan images for each exposure condition for all samples on the chip into a single Excel file.
4. For each sample, select the optimal exposure condition such that the net intensity for all lectin spots is <40,000 (*see* **Note 36**).
5. For the selected data, calculate the relative signal intensities of 45 lectins (*see* **Note 37**), normalizing each value to the mean intensity across all lectins. The resulting mean normalized values represented the glycomic profiles of the tissue samples.
6. Prior to the statistical analysis, the raw datasets should be qualified. Samples with net intensities that differ substantially (e.g., Fig. 6a, samples 1 and 2) should be excluded, as such

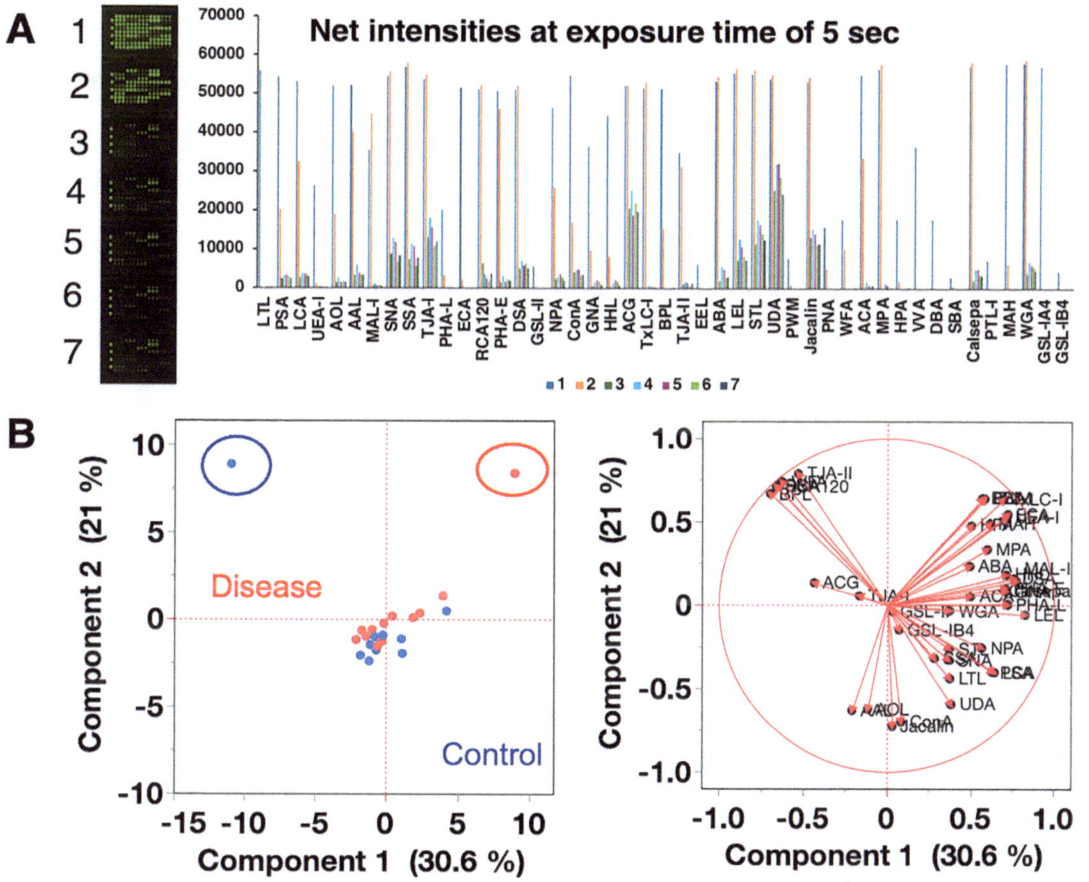

Fig. 6 Data quality assessment and qualification. (**a**) Example of raw data exhibiting excessive signal intensities due to technical errors or unequal sample loading. Samples with signals (No. 1–2) that deviate significantly from others should be excluded before statistical analysis. (**b**) Example of PCA plot for disease and control samples used for data qualification. Samples marked by circles are considered outliers

deviations may result from technical errors or excessive protein loading. Repeat the experiment with a smaller sample volume to confirm whether the difference was biologically significant. If the results remain unchanged, it can be considered genuine. Another approach for data qualification is to use a principal component analysis (PCA) plot to identify and exclude samples with profiles that differ markedly from the rest (Fig. 6b).

7. Figure 7 shows the average glycomic profiles of the heart disease WFA^{+} (Fig. 4c, d) and WFA^{-} (Fig. 4e, f) groups within the same tissue samples (n = 3 regions [3 slide sections]), indicating that reproducible data can be obtained using this method.

8. Figure 8 shows the lectins with statistically significant differences in the heart disease tissue between the WFA^{+} and WFA^{-} groups (A), and between the disease and control groups (B), as

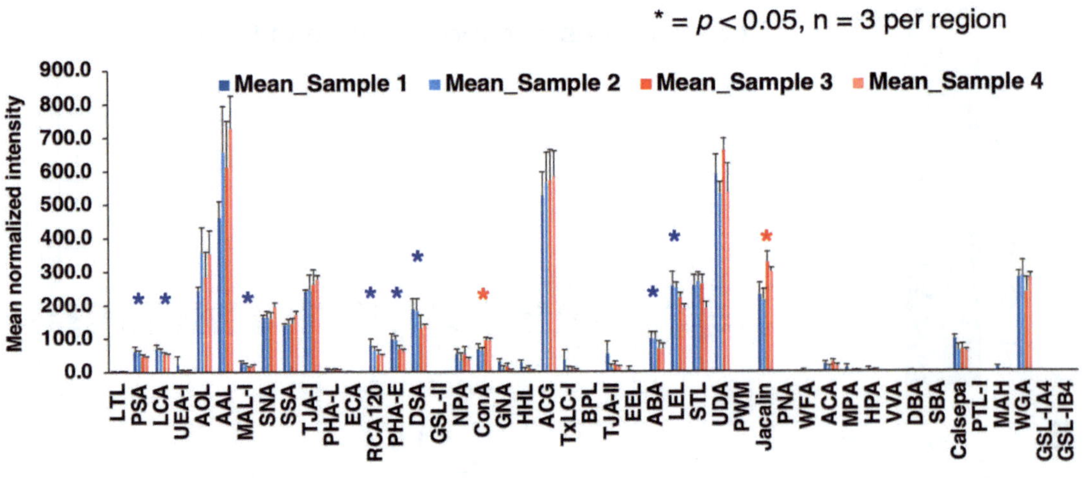

Fig. 7 LMD-LMA data reproducibility. Bar chart depicting the average mean-normalized intensity of 45 lectins in WFA^{+} and WFA^{-} regions from triplicate tissue sections of fibrotic heart tissues measured using Method 2. Blue and red asterisks (*) indicate lectins with significantly increased and significantly decreased expression in WFA^{+} regions, respectively

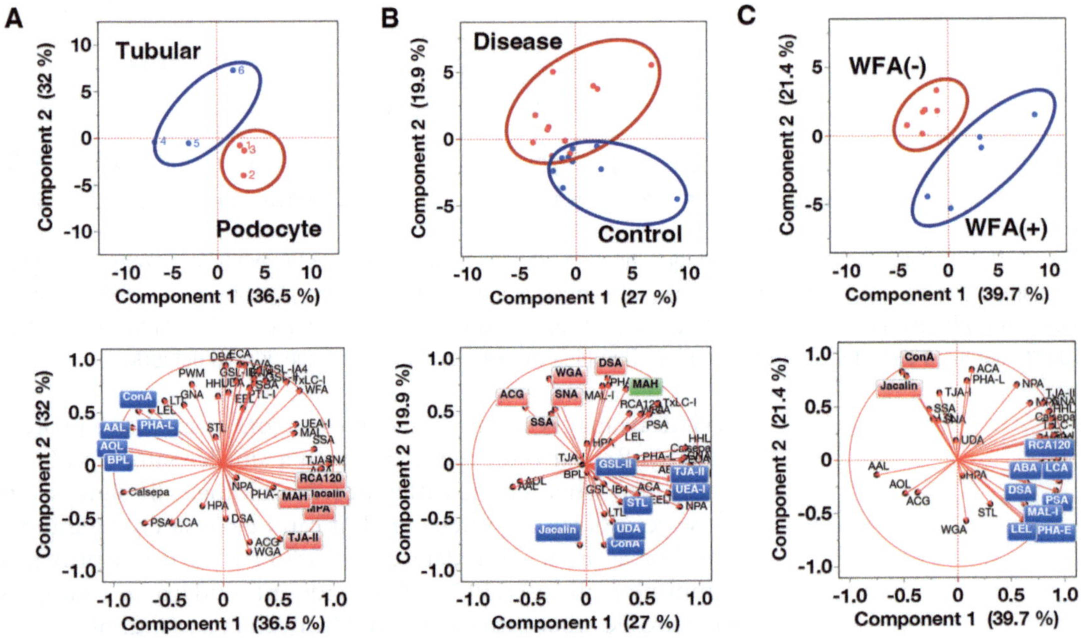

Fig. 8 Statistical comparison of glycomics profiles. Principal component analysis (PCA) was performed for the following datasets: (**a**) comparison of mouse renal tissue compartments (glomeruli versus tubules) within the same tissue section using Method 1. (**b**) Comparison of rat hearts of disease and control groups using Method 2. (**c**) WFA^{+} (fibrotic) and WFA^{-} (non-fibrotic) rat heart tissues using Method 2. In the PCA loading plots, blue and red markers represent lectins showing statistically significant increases ($p < 0.05$), corresponding to the groupings shown in the PCA score plots above. MAH lectin (green) indicates an increase with $\log_2$ fold change >2

well as differences in the glycomic profiles of kidney tubules and glomeruli (C). These results demonstrate the feasibility of the LMD-LMA method for distinguishing intra-tissue glycomic profiles by targeted sampling and inter-tissue differences between diseased and normal groups.

4 Notes

1. Prepare FFPE tissue blocks using similar fixation conditions (same fixative and incubation time) for all tissues to be compared because these parameters can affect glycomic profiles. For the protocol for FFPE mouse tissue preparation, please see references [3, 15, 19].
2. Use an antigen retrieval buffer at pH 6.0, similar to that used in LMD-LMA, for histochemical analysis to evaluate targeted tissue glycome mapping. This is because the LMD-LMA method contains a heat denaturation step corresponding to antigen retrieval (Subheading 3.4, **Step 2**), which may influence both the net quantity of lectin-binding glycoproteins and the lectin staining pattern.
3. Streptavidin labeled with other dyes with different fluorescence wavelengths that are compatible with the filters of a fluorescence microscope can also be used. Accordingly, a preliminary experiment may be useful to determine the optimal dye for the fluorescence staining of tissue sections, considering the autofluorescence from the tissue section to be analyzed.
4. Prepare a sufficient volume of solution for all samples and negative controls, along with one extra reserve.
5. Use wide-tip pipettes for NP-40 because of its high viscosity. Similarly, avoid volume reduction during pipetting as this may lead to inaccurate measurements.
6. Use the specified high-power sonicator to ensure efficient sonication.
7. Use low-retention tips and tubes to handle glycoprotein samples because their small volume increases the risk of protein loss during preparation and LMA analysis.
8. Ensure that the tubes are properly sealed to prevent moisture ingress, which causes Cy3-SE degradation and a short storage life.
9. Use the same lot of lectin array chips for the whole analysis. The use of chips from different lots may result in binding activity that can lead to inter-batch variability. LecChip Ver.1.0 is discontinued. The successor product, LecChip uni-45, is available from Precision System Science Co., Ltd.

10. This step prevents tissue sections from detaching from the glass slides during subsequent procedures.
11. The immersion time required for proper staining varies according to the tissue type. Therefore, start with a short staining time (1–3 minutes), especially when using freshly prepared dye, and check quickly under a microscope to determine whether more staining is required. Notably, the staining will become stronger after drying.
12. Please ensure that the slide is completely dry as any remaining water may lead to a failed dissection.
13. Proceed to LMD and protein extraction steps within 1 week of staining to minimize the effects of potential tissue protein degradation on glycomic profiling.
14. Preparing an additional serial section stained with hematoxylin and eosin, which enables detailed morphological observation, can help facilitate the identification of target areas for dissection.
15. Because most commercial lectins and antibodies are glycoproteins, their attached glycans can affect glycomic profiling. Therefore, if necessary, assess the contamination level of the biotinylated probe using a Western blot-like chemiluminescence detection system with peroxidase-conjugated streptavidin, and evaluate its impact on the resulting glycomic profiles.
16. Rapid temperature changes should be avoided because they may cause detachment of the tissue section from the glass slide. Perform all procedures quickly after the first addition of the solution to prevent the section from drying.
17. OPTIONAL: Use a rotator shaker for all incubation steps hereafter to ensure that equilibrium is reached in the indicated times.
18. To ensure efficient tissue collection, ensure that dissected fragments fall directly into the collection tube caps. To prevent static electricity, use an antistatic sheet or ionizer on slides and tubes before use. For the same reason, handle slides and tubes with bare hands during dissection of noninfectious tissue samples.
19. During tissue dissection, capture images at the following three time points for each dissected area: (1) before dissection without frames, (2) after drawing the frames of the dissected areas, and (3) after dissection. These images enable a detailed analysis of the morphology and cellular composition within the dissected regions.
20. Set the laser power to the lowest level to minimize tissue burnout during dissection. As the optimal power varies by tissue type, determine the best settings through a preliminary

study. High cutting speeds often result in incomplete dissection, whereas lower speeds allow gravity to pull the fragment away from the slide so that it falls directly into the collection tube cap.

21. Since the dissected and fluorescently stained sections are serial sections on different slides, it is advisable to clearly identify and mark the target areas before dissection.
22. Ensure that the tissue fragments are collected at the bottom of the tube in this step as fragments stuck to the tube walls are difficult to pellet by centrifugation after the addition of the solution.
23. Slowly add the buffer to prevent tissue fragments from floating. If the fragments float, gently press them back down using a pipette tip. Floating fragments may lead to incomplete antigen retrieval.
24. Carefully aspirate the supernatant without disturbing the pellet to avoid the loss of tissue fragments. For this reason, the aspirated volume is set at 190 μL.
25. Adding 10 μL of PBS containing 1% NP-40 to the tissue fragment suspension results in a final protein extraction buffer with 0.5% NP-40 concentration.
26. Ensure that each tube is adequately exposed to ultrasonic waves as this step critically affects protein extraction efficiency. It is recommended that the tube be chilled on ice between sonication intervals to prevent overheating.
27. Carefully collect the supernatant without pipetting tissue fragments as contamination may cause increased noise or unexpected spots in the LMA analysis.
28. Note that unlabeled protein samples can be stored long term at −30 °C; however, long-term storage of Cy3-labeled protein samples may lead to decreased signals in the LMA analysis.
29. Use working Cy3-SE dye prepared from the same batch and at the same time for all protein samples to be compared, because different dye lots can vary in labeling efficiency and affect glycomic profiles. Ensure that the Cy3-SE dye dissolves completely when it is used.
30. Use lectin array chips from the same or consecutive lot numbers with consistent properties for all protein samples to be compared. Because chips are manufactured in lots of 10, interlot variation should be assessed in a preliminary study when processing more than 70 samples.
31. To minimize the carryover of liquids, all washing steps should be performed as follows. Discard the solution by turning the chip face down and gently swinging the chip on a liquid waste

bottle, then place the chip face down on a polypropylene nonwoven wiper to absorb excess liquids from the wells without touching the surface of the wells. Turn the chip face up and immediately add the desired solution. Perform all procedures quickly after the first addition of the solution to the chip to avoid drying the wells. For this reason, the use of the 100–1000 μL auto-dispensing pipette for washing steps is recommended.

32. For glycomic profiling, the amount of protein sample should be optimized to avoid saturation of glycoprotein binding across all lectin spots. Determine this optimal amount through a preliminary study using serial dilutions of the test samples. Dilute Cy3-labeled protein samples with probing buffer as needed.
33. Avoid touching the sides of the chip and carefully clean the sides and bottom of the chip with Kimwipes. Because the scanner operates horizontally, contamination, such as fingerprints and dried probing buffer salts, can cause blurred or high-background images, leading to inaccurate analysis.
34. The scanning time and binning mode can be adjusted to match the amount of loaded glycoprotein. For very small sample amounts, 2 × 2 and 4 × 4 binning can be used to enhance signal detection.
35. The net intensity of each lectin spot is automatically calculated by subtracting the background, defined as the mean signal of the four background grids, from the signal intensity (Fig. 2). The net intensity of each lectin was determined as the average of three spot signals. Ensure that all lectin signals exhibit similar standard deviations (S.D.); if a relatively large S.D. is observed, verify that the lectin spot grids are correctly aligned.
36. Because the limit value of the net intensity is set to 67% of the maximum value to select the gain condition with unsaturated signals, set the limit value according to the performance of the scanner.
37. The GlycoStation Tool Pro ver. 1.5 includes 45 lectins as it was developed for the LecChip ver. 1.0. However, this software allows for the exclusion of unwanted lectins.

Acknowledgments

This study was partially supported by JSPS KAKENHI grants (JP24K19073 to P.B. and JP20K17136 and JP24K11257 to C. N.-O).

References

1. Zhang H, Lu KH, Ebbini M et al (2024) Mass spectrometry imaging for spatially resolved multi-omics molecular mapping. NPJ Imaging 2:20
2. Guo Y, Wang W, Ye K et al (2023) Single-nucleus RNA-Seq: open the era of great navigation for FFPE tissue. Int J Mol Sci 24:13744
3. Mennillo E, Kim YJ, Lee G et al (2024) Single-cell and spatial multi-omics highlight effects of anti-integrin therapy across cellular compartments in ulcerative colitis. Nat Commun 15: 1493
4. Wallace EN, West CA, Mcdowell CT et al (2024) An N-glycome tissue atlas of 15 human normal and cancer tissue types determined by MALDI-imaging mass spectrometry. Sci Rep 14:489
5. Isnard P, Humphreys BD (2025) Spatial transcriptomics: integrating morphology and molecular mechanisms of kidney diseases. Am J Pathol 195:23–39
6. Espina V, Wulfkuhle JD, Calvert VS et al (2006) Laser-capture microdissection. Nat Protoc 1:586–603
7. Mund A, Coscia F, Kriston A et al (2022) Deep visual proteomics defines single-cell identity and heterogeneity. Nat Biotechnol 40:1231–1240
8. Varki A, Kornfeld S (2022) Historical background and overview. In: Varki A, Cummings RD, Esko JD, Stanley P, Hart GW, Aebi M, Mohnen D, Kinoshita T, Packer NH, Prestegard JH, Schnaar RL, Seeberger PH (eds) Essentials of glycobiology. Cold Spring Harbor Laboratory Press, Cold Spring Harbor
9. Nagai-Okatani C, Nishigori M, Sato T et al (2019) Wisteria floribunda agglutinin staining for the quantitative assessment of cardiac fibrogenic activity in a mouse model of dilated cardiomyopathy. Lab Investig 99:1749–1765
10. Pienkowski T, Wawrzak-Pienkowska K, Tankiewicz-Kwedlo A et al (2025) Leveraging glycosylation for early detection and therapeutic target discovery in pancreatic cancer. Cell Death Dis 16:227
11. Tanaka-Okamoto M, Hanzawa K, Yamamoto T et al (2025) Combination of urinary free-glycan markers for the diagnosis of various malignant tumors. Sci Rep 15:10109
12. Haslam SM, Freedberg DI, Mulloy B et al (2022) Structural analysis of glycans. In: Varki J, Cummings RD, Esko JD, Stanley P, Hart GW, Aebi M, Mohnen D, Kinoshita T, Packer NH, Prestegard JH, Schnaar RL, Seeberger PH (eds) Essentials of glycobiology. Cold Spring Harbor. Cold Spring Harbor Laboratory Press
13. Boottanun P, Fuseya S, Kuno A (2025) Toward spatial glycomics and glycoproteomics: innovations and applications. BBA Adv 7:100146
14. Kuno A, Uchiyama N, Koseki-Kuno S et al (2005) Evanescent-field fluorescence-assisted lectin microarray: a new strategy for glycan profiling. Nat Methods 2:851–856
15. Boottanun P, Nagai-Okatani C, Nagai M et al (2023) An improved evanescent fluorescence scanner suitable for high-resolution glycome mapping of formalin-fixed paraffin-embedded tissue sections. Anal Bioanal Chem 415:6975–6984
16. Zou X, Yoshida M, Nagai-Okatani C et al (2017) A standardized method for lectin microarray-based tissue glycome mapping. Sci Rep 7:43560
17. Nagai-Okatani C, Nagai M, Sato T et al (2019) An improved method for cell type-selective glycomic analysis of tissue sections assisted by fluorescence laser microdissection. Int J Mol Sci 20:700
18. Nagai-Okatani C, Aoki-Kinoshita KF, Kakuda S et al (2019) LM-GlycomeAtlas ver. 1.0: a novel visualization tool for lectin microarray-based glycomic profiles of mouse tissue sections. Molecules 24:2962
19. Nagai-Okatani C, Zou X, Fujita N et al (2021) LM-GlycomeAtlas ver. 2.0: an integrated visualization for lectin microarray-based mouse tissue glycome mapping data with lectin histochemistry. J Proteome Res 20:2069–2075
20. Nagai-Okatani C, Fujita N, Boottanun P et al (2025) LM-GlycoRepo version 1.0: a novel repository system for mouse tissue glycome mapping data. J Proteome Res 24:3569–3577
21. Nagai-Okatani C, Zou X, Matsuda A et al (2022) Tissue glycome mapping: lectin microarray-based differential glycomic analysis of formalin-fixed paraffin-embedded tissue sections. Methods Mol Biol 2460:161–180

Chapter 4

Laser Capture Microdissection Followed by DNA Methylation Profiling

Lu Liu, Yingchuo Hu, Chunhong Zheng, and Yanyi Huang

Abstract

Laser capture microdissection (LCM) enables the precise isolation of morphologically defined cell populations, overcoming the confounding effects of tissue heterogeneity in DNA methylation analyses. However, the low DNA yield inherent to LCM—often derived from fewer than 1000 cells—challenges conventional bisulfite-based sequencing, which requires high DNA input and induces severe degradation. Here, we integrate LCM with Enzymatic Methyl-seq (EM-seq), a bisulfite-free, enzymatic conversion method that preserves DNA integrity by minimizing damage and reducing GC bias. Our optimized low-input library preparation protocols allow robust and reproducible methylation profiling from as few as 100 cells, with data quality comparable to that of bulk tissue samples. This strategy not only enables the isolation of tumor cells to uncover methylation signatures for early biomarker discovery and targeted therapies in oncology but also holds promise for advancing personalized medicine through patient-specific epigenetic profiling. Overall, our work establishes LCM-EM-seq as a scalable framework for spatially resolved, methylation-aware analyses of rare cell populations, opening new avenues for studying epigenetic heterogeneity in development and disease.

Key words DNA methylation profiling, Spatially resolved methylation analysis, Low-input library preparation, Bisulfite-free conversion, Enzymatic Methyl-seq

1 Introduction

DNA methylation is a fundamental epigenetic mechanism that regulates gene expression in a cell-type-specific manner [1, 2], yet its study in bulk tissues can be confounded by cellular heterogeneity [3]. Laser capture microdissection (LCM) addresses this challenge by enabling the precise isolation of morphologically distinct cell populations directly from tissue sections [4]. However, LCM-derived samples typically yield low quantities of DNA, often from fewer than 1000 cells [5], which poses challenges for conventional bisulfite sequencing (BS-seq) methods [6, 7]. BS-seq requires substantial input DNA and involves harsh treatment con-

Roberta Noberini (ed.), *Laser Capture Microdissection: Methods and Protocols*, Methods in Molecular Biology, vol. 3015,
https://doi.org/10.1007/978-1-0716-5154-4_4,

ditions that can degrade over 90% of the DNA [8], leading to issues such as GC bias [9] and reduced resolution in low-input scenarios [10, 11].

Recent enzymatic conversion methods, such as Enzymatic Methyl-seq (EM-seq) [12], offer a compelling alternative. By leveraging enzymatic cytosine deamination, EM-seq preserves DNA integrity and provides accurate methylation profiling from subnanogram inputs while detecting both 5-methylcytosine and 5-hydroxymethylcytosine [12]. This method minimizes issues like GC bias and uneven coverage [13], making it particularly suitable for LCM-derived samples.

The integration of LCM with EM-seq has broad applications in biological and medical research [14]. In oncology, this strategy could facilitate the isolation of tumor cells to uncover methylation signatures that could serve as early biomarkers or guide targeted therapies [15, 16]. In neuroscience, it could enable the examination of epigenetic modifications in discrete neuronal populations, shedding light on the molecular underpinnings of neurodegenerative and psychiatric disorders [17–19]. Moreover, in developmental biology, it could support investigations into the dynamic methylation changes that drive tissue differentiation and organogenesis [20, 21], while in clinical settings, it holds promise for advancing personalized medicine through patient-specific epigenetic profiling [22].

In this chapter, we present an integrated approach combining LCM with EM-seq, coupled with optimized low-input library preparation techniques, to achieve high-resolution methylation profiling of small cell populations (Fig. 1). We tested cell inputs ranging from 100 to 1000 cells, and all yielded data of comparable quality to that obtained from traditional bulk samples. This approach preserves both spatial and epigenetic fidelity while effectively overcoming the limitations inherent in conventional BS-seq, thus establishing a robust workflow for exploring epigenetic heterogeneity across diverse biological contexts.

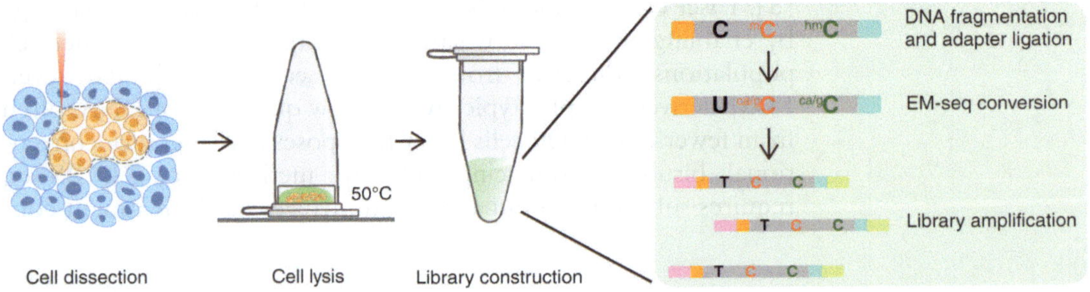

Fig. 1 Overview of the whole-genome methylation library construction protocol following LCM. Cell clusters are first dissected, followed by cell lysis, and libraries are then constructed using the EM-seq method

2 Materials

2.1 Frozen Tissue Sectioning and Staining

1. Optimal cutting temperature (OCT) medium.
2. Mayer's hematoxylin solution.
3. 1% eosin solution.
4. Polyethylene naphthalate (PEN) membrane slides.
5. Microscope slides.
6. Distilled water.
7. 100% ethanol.
8. 70% ethanol.
9. 95% ethanol.
10. Cryostat.
11. Light microscope.

2.2 Laser Capture Microdissection and Cell Lysis

1. 0.2 mL PCR tubes.
2. Laser microdissection microscope.
3. Cell lysis buffer: 4‰ Triton X-100, 400 μg/mL proteinase K, 30 mM Tris–HCl (pH = 7.6), 10 mM NaCl, 5 mM EDTA.

2.3 Whole-Genome Methylation Library Construction

1. NEBNext Enzymatic Methyl-seq Kit, including NEBNext Ultra II End Prep Enzyme Mix, NEBNext Ultra II Ligation Master Mix, NEBNext Ligation Enhancer, TET2 Reaction Buffer, TET2 Reaction Buffer Supplement, 100 mM DTT, Oxidation Supplement, Oxidation Enhancer, TET2, Fe (II) Solution, Stop Reagent, APOBEC, APOBEC Reaction Buffer, BSA, NEBNext Q5U Master Mix, NEBNext EM-seq Adapter, EM-seq Index Primer, and NEBNext Sample Purification Beads.
2. Spike-in mix: 1 pg/μL CpG methylated pUC19 control DNA, 20 pg/μL CpG unmethylated Lambda control DNA.
3. Fragmentation mix, made from components of NEBNext UltraShear (NEB): protease inhibitors (e.g., 1.6× Promega protease inhibitor cocktail), 1 μL spike-in mix, 1.47 mM $MgCl_2$, 14 μL NEBNext UltraShear Reaction Buffer, 4 μL NEBNext UltraShear. Prepare a total volume of 34 μL for each sample (*see* **Note 1**).
4. 500 mM DTT.
5. Diluted methyl adapter: Make a two-fold dilution of the NEBNext EM-seq Adapter in nuclease-free water (*see* **Note 2**).
6. Magnetic stand.
7. 80% ethanol.

8. TET2 buffer: Add 100 μL of TET2 Reaction Buffer to one tube of TET2 Reaction Buffer Supplement and mix well (*see* **Note 3**). The reconstituted buffer should be stored at −20 °C and discarded after 4 months.
9. TET2 mix: 10 μL TET2 buffer, 1 μL Oxidation Supplement, 1 μL 100 mM DTT, 1 μL Oxidation Enhancer, 4 μL TET2.
10. Diluted Fe(II) Solution: Dilute the Fe(II) Solution by adding 1 μL to 1249 μL of nuclease-free water.
11. 100% formamide.
12. APOBEC mix: 1 μL APOBEC, 1 μL BSA, 10 μL APOBEC Reaction Buffer, 68 μL nuclease-free water.
13. PCR mix: 25 μL NEBNext Q5U Master Mix, 5 μL EM-seq Index Primer.
14. Qubit 1X dsDNA HS Assay kit.
15. Agilent Fragment Analyzer System.
16. Illumina 2 × 150 bp paired-end sequencing kit.

3 Methods

Maintain samples on ice throughout all steps unless explicitly noted. Employ low-binding tubes and filtered tips to minimize sample adhesion during liquid handling.

3.1 Frozen Tissue Sectioning and Staining

1. Embed freshly frozen tissues in OCT medium at −25 °C.
2. Section a 3 μm thick tissue slice using a Leica Cryostat on a microscope slide. Stain the section using a standard hematoxylin and eosin (H&E) staining procedure and then analyze it under light microscopy. Use this slice as a morphological and histological reference during LCM.
3. Obtain 10 μm thick adjacent sections on PEN membrane slides. Immerse slides in 70% ethanol for 15 s. Transfer slides to double-distilled water for 30 s to remove residual ethanol.
4. Stain nuclei by submerging slides in hematoxylin solution for 50 s. Rinse slides under low-flow running tap water for 30 s to remove excess stain.
5. Counterstain the cytoplasm by dipping the slides in the eosin solution for 3–5 s. Rinse the slides gently under low-flow running tap water for 1–1.5 min to clear residual eosin.
6. Dehydrate the slides by first dipping them in 70% ethanol briefly for 3–5 s. Transfer the slides to 95% ethanol for 3–5 s. Complete the dehydration in 100% ethanol for 3–5 s.

3.2 Laser Capture Microdissection and Cell Lysis

1. For each cell sample, freshly prepare 10 μL of cell lysis buffer before LCM.
2. Perform LCM under real-time microscopic visualization. Use a 5× objective to locate target tissue regions, then switch to a 20× objective to identify and dissect specific cell clusters (approximately 100 cells per sample in this example protocol) to PCR tube caps.
3. Gently pipette 10 μL of cell lysis buffer directly onto the PCR tube cap containing the dissected cell clusters. Securely close the tube in an upside-down position to retain the buffer within the cap during incubation.
4. Incubate the tubes at 50 °C for 3 h in a thermostatic incubator or metal bath.
5. After lysis, briefly vortex the tubes and centrifuge it to collect the lysates (*see* **Note 4**). The lysed DNA can be stored at −20 °C overnight or − 80 °C for up to a month.

3.3 Whole-Genome Methylation Library Construction

3.3.1 DNA Fragmentation, End Preparation, and Adapter Ligation

1. Prepare the fragmentation mix on ice and combine 34 μL fragmentation mix to the 10 μL of DNA lysates. Vortex for 5 s and briefly spin.
2. Incubate at 37 °C for 15 min and at 65 °C for 15 min to inactivate the endonuclease (*see* **Note 5**). Hold at 4 °C.
3. Add 2 μL of 500 mM DTT and 3 μL NEBNext Ultra II End Prep Enzyme Mix to the fragmented DNA. Mix thoroughly by vortexing for 2 s and briefly spin.
4. Incubate at 20 °C for 15 min and at 65 °C for 15 min, hold at 4 °C.
5. Add 2.5 μL of diluted methyl adapter to the end preparation reaction mixture. Mix thoroughly by vortexing for 2 s and briefly spin. Add 1 μL NEBNext Ligation Enhancer and 30 μL NEBNext Ultra II Ligation Master Mix to the diluted methyl adapter-containing end preparation reaction mixture. Mix thoroughly by vortexing for 2 s and briefly spin.
6. Incubate at 20 °C for 20 min, hold at 4 °C (*see* **Note 6**).
7. Resuspend NEBNext Sample Purification Beads by vortexing. Add 93 μl of resuspended beads to each sample. Mix by vortexing and briefly spin. Incubate at room temperature for 5 min.
8. Put the tubes in a magnetic stand, wait for 5 min or until the supernatant is clear.
9. Remove the supernatant thoroughly without disturbing the beads.
10. Add 200 μL of 80% ethanol to each sample while the tubes are still on the magnetic stand. Wait for 30 s, then remove the supernatant thoroughly.

11. Repeat for a total of two washes. Remove the ethanol completely.
12. Keep the lids open while the tubes are on the magnetic stand. Allow the beads to dry in air for 5–10 min (*see* **Note 7**).
13. When the first crack appears on the surface of the beads, remove the tubes from the magnetic stand and thoroughly resuspend the magnetic beads either by vortex or pipetting in 30 μL of H_2O per sample.
14. Incubate the tubes at room temperature for 5 min. Transfer the tubes to the magnetic stand, wait for 5 min or until the supernatant is clear.
15. Transfer 28 μL of supernatant to a new PCR tube per sample carefully without disturbing the bead pellets. Discard the beads.

3.3.2 Oxidation of 5-mC and 5-hmC

1. Add 17 μL of TET2 mix to the 28 μL purified DNA fragments. Mix thoroughly by vortexing and spin briefly.
2. Combine 5 μL of diluted Fe(II) Solution and 45 μL TET2 reaction mixture containing DNA fragments. Mix thoroughly by vortexing and spin briefly. Incubate at 37 °C for 1 h.
3. Add 1 μl of Stop Reagent. Mix thoroughly by vortexing and spin briefly. Incubate at 37 °C for 30 min, hold at 4 °C.
4. Perform DNA purification by adding 90 μL of NEBNext Sample Purification Beads to each sample. Follow the same procedures as introduced in Subheading 3.3.1. Resuspend air-dried beads in 18 μL H_2O and transfer 16 μL of eluate to a new tube.

3.3.3 Deamination of Cytosines

1. Preheat a thermocycler to 85 °C with the heated lid on.
2. Add 4 μl formamide to the 16 μl of oxidized DNA. Vortex to mix and spin briefly.
3. Incubate at 85 °C for 10 minutes in the preheated thermocycler with the heated lid on.
4. Immediately place on ice and allow the sample to fully cool (~2 minutes) before proceeding to the next step.
5. Add 80 μl APOBEC mix to the 20 μL of denatured DNA.
6. Mix thoroughly by vortexing and spin briefly.
7. Incubate at 37 °C for 3 hours, then at 4 °C in a thermocycler with the heated lid on.
8. Perform DNA purification by adding 100 μL of NEBNext Sample Purification Beads to each sample. Follow the same procedures introduced in Subheading 3.3.1. Resuspend air-dried beads in 22 μL H_2O and transfer 20 μL of eluate to a new tube.

3.3.4 Library Amplification and Quantification

1. Add 30 μL of PCR mix to the 20 μL of purified DNA fragments. Mix thoroughly by vortexing and spin briefly.
2. Place the tube in a thermocycler and perform 12 cycles of PCR amplification (*see* **Note 8**) using the following cycling conditions:
   - Denaturation: 98 °C, 30 s.
   - 12 cycles: 98 °C, 10 s; 62 °C, 30 s; 65 °C, 60 s.
   - Final extension: 65 °C, 5 min.
   - Hold at 4 °C.
3. Perform DNA purification by adding 40 μL of NEBNext Sample Purification Beads to each sample. Follow the same procedures introduced in Subheading 3.3.1. Resuspend air-dried beads in 22 μL of H_2O and transfer 20 μL of eluate to a new tube.
4. Quantify the concentration of the libraries using the Qubit 1X dsDNA HS Assay kit (*see* **Note 9**).
5. Determine the distribution of library fragments using an Agilent Fragment Analyzer System (Fig. 2; *see* **Note 10**). Proceed with Illumina sequencing with the 2 x 150 bp paired-end sequencing kit.

3.4 Data Analysis

We use a Linux command-line interface to process whole-genome methylation sequencing data. We recommend using Conda [23] (https://github.com/conda/conda) to create a virtual environment for software installation and data processing.

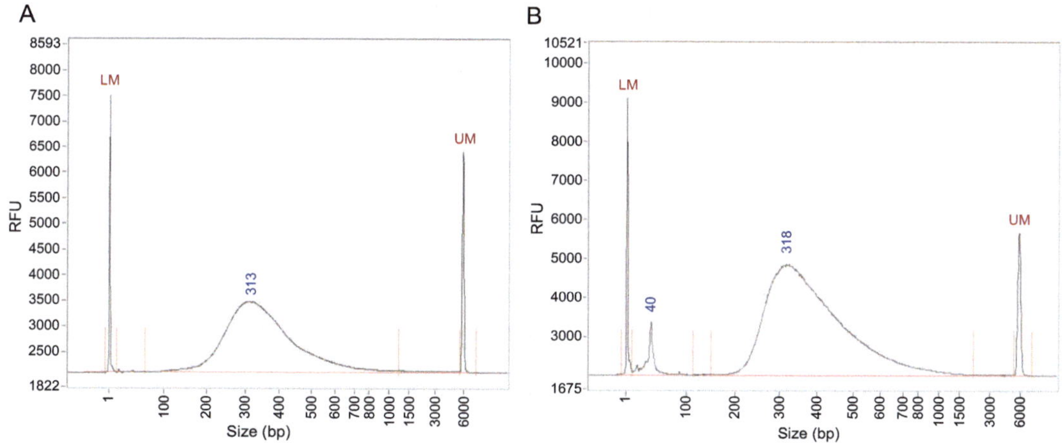

Fig. 2 Representative Fragment Analyzer electropherograms of the LCM-derived methylation libraries constructed from 200 human oral squamous cell carcinoma cells [28]. RFU, relative fluorescence units; LM, lower marker; UM, upper marker. **(a)** The library exhibits a fragment size distribution ranging from 200 to 800 bp. **(b)** Occasionally, residual PCR primer peaks are detected in the library (*see* **Note 11**)

3.4.1 Preparation of the Analysis Environment

1. Create environment:

```
conda create -n methyl python = 3.10
conda activate methyl
```

2. Install required software including Cutadapt [24] (https://github.com/marcelm/cutadapt), Bismark [25] (https://github.com/FelixKrueger/Bismark), Biowtie2 [26] (https://github.com/BenLangmead/bowtie2), and samtools [27] (https://github.com/samtools/samtools):

```
conda install -c bioconda \
     cutadapt=4.2 \
     bismark=0.24.1 \
     bowtie2=2.5.1 \
     samtools=1.18
```

3. Download corresponding reference genome, for example, the human reference genome hg38: https://hgdownload.soe.ucsc.edu/goldenPath/hg38/bigZips/latest/hg38.fa.gz).
4. Build a Bismark index by running:

```
  bismark_genome_preparation \
    --bowtie2 \
    --path_to_aligner /absolute/path/to/bowtie2/bin/bowtie2-
build \
    --verbose \
    --parallel 8 \
    /absolute/path/to/genomes/hg38_directory
```

3.4.2 Methylation Alignment and Extraction

1. Perform adapter trimming with Cutadapt:

```
cutadapt \
     --minimum-length 20 \
     -q 20\
     --pair-filter=both\
     -a AGATCGGAAGAGCACACGTCTGAACTCCAGTCA \
     -A AGATCGGAAGAGCGTCGTGTAGGGAAAGAGTGT \
     -o /absolute/path/to/trimmed/fastq/trimmed_sample_1.fq.
gz \
     -p /absolute/path/to/trimmed/fastq/trimmed_sample_2.fq.
gz \
     /absolute/path/to/raw/fastq/raw_sample_1.fq.gz \
     /absolute/path/to/raw/fastq/raw_sample_2.fq.gz \
```

2. Bisulfite alignment with Bismark:

```
bismark \
     --bowtie2 \
     --genome /absolute/path/to/hg38_genome \
     --multicore 4 \
```

```
        -p 4\
        --un \
        --temp_dir /absolute/path/to/temp_files \
       -1 /absolute/path/to/trimmed/fastq/trimmed_sample_1.fq.
gz \
       -2 /absolute/path/to/trimmed/fastq/trimmed_sample_2.fq.
gz \
        -o /absolute/path/to/results/experiment_sample
```

3. Remove PCR duplicate reads:

```
   deduplicate_bismark \
           --bam /absolute/path/to/results/experiment_sample/
experiment_sample_pe.bam \
        --output_dir /absolute/path/to/results/experiment_sam-
ple
```

4. Methylation extraction:

```
   bismark_methylation_extractor \
       --gzip \
       --bedGraph \
       --cytosine_report \
       --genome_folder /absolute/path/to/hg38_genome \
       --parallel 4 \
      -o /absolute/path/to/results/experiment_sample \
     /absolute/path/to/results/experiment_sample/experiment_-
sample_pe.deduplicated.bam
```

5. Perform nucleotide coverage analysis:

```
   bam2nuc \
       --dir /absolute/path/to/results/experiment_sample \
       --genome_folder /absolute/path/to/hg38_genome \
      --samtools_path /absolute/path/to/samtools \
     /absolute/path/to/results/experiment_sample/experiment_-
sample_pe.deduplicated.bam
```

3.4.3 Methylation Statistics and Quality Controls (QCs)

1. Generate methylation QC report:

```
   bismark2report \
       --dir /absolute/path/to/results/experiment_sample \
         --alignment_report /absolute/path/to/results/experi-
ment_sample/experiment_sample_PE
      _report.txt \
     --dedup_report /absolute/path/to/results/experiment_sam-
ple/experiment_sample_
```

```
deduplication_report.txt \
   --splitting_report /absolute/path/to/results/experi-
ment_sample/experiment_sample_
splitting_report.txt \
--mbias_report /absolute/path/to/results/experiment_sam-
ple/experiment_sample.M-
bias.txt \
  --nucleotide_report /absolute/path/to/results/experi-
ment_sample/experiment_sample.
nucleotide_stats.txt
```

2. To validate the conversion efficiency, download the genome references for lambda (NCBI accession: J02459) and pUC19 (L09137), and build the bismark indices for both genomes. Align unmapped reads from the hg38 alignment step to the lambda and pUC19 references using the same parameters and workflow as for the human genome (*see* **Note 12**).

4 Notes

1. Vortex the NEBNext UltraShear briefly prior to use for optimal performance.
2. The appropriate dilution factor for the NEBNext EM-seq Adapter depends on the number of input cells. Based on our experience, a two-fold dilution is recommended for a cell input ranging from 100 to 500. For inputs exceeding 500 cells, no dilution is necessary.
3. Centrifuge the lyophilized TET2 Reaction Buffer Supplement before use to ensure that it is fully collected at the bottom of the tube.
4. We tested cell inputs ranging from 100 to 1000 cells; within this range, the cell lysate can be used directly for DNA fragmentation. For inputs exceeding 1000 cells, an additional genomic DNA purification step may be required to prevent cellular contaminants from interfering with downstream reactions. Commercial genomic DNA purification kits can be used for this purpose. When using purified genomic DNA, protease inhibitors and $MgCl_2$ should be omitted from the fragmentation mix.
5. The incubation time at 37 °C should be determined based on the desired fragmentation profile, which may vary depending on the sample type (e.g., degraded DNA from samples that have been frozen for an extended period). Optimization is required to achieve an optimal fragmentation, and the recommended fragmentation times provided by NEB can serve as a guideline.

6. To enhance ligation efficiency—especially for low-input samples—consider extending the ligation reaction time to up to 4 hours or overnight at 4 °C; however, note that excessively long ligation times may increase the risk of adapter self-ligation.
7. Overdrying the beads can lead to low DNA recovery. Perform the elution step as soon as all visible liquid has evaporated, avoiding overdrying.
8. The optimal number of PCR cycles depends on the input cell quantity. For example, we recommend 12 cycles for a 100-cell input, 10 cycles for 500 cells, and 8 cycles for 1000 cells.
9. The final yield of each library may vary from 100 to 500 ng.
10. Alternatively, fragment distribution can be determined using any equivalent platform that generates detailed electropherogram profiles, allowing for assessment of fragment size range, peak intensity, and overall library quality.
11. Perform an additional purification step using 0.8× purification beads to remove residual PCR primers.
12. Unmethylated lambda DNA serves as negative control (expected 0% CpG methylation) and pUC19 DNA acts as positive control (expected 100% CpG methylation). Methylation in all contexts (CpG, CHG, CHH) is typically <0.6% for unmethylated lambda DNA, around 96% for pUC19 CpG methylation, and <1.6% in the CHG and CHH contexts. This confirms high enzymatic conversion efficiency, with minimal nonspecific cytosine deamination in the negative control and near-complete retention of methylated cytosines in the positive control.

References

1. Bird A (2002) DNA methylation patterns and epigenetic memory. Genes Dev 16:6–21
2. Ziller MJ, Gu H, Müller F et al (2013) Charting a dynamic DNA methylation landscape of the human genome. Nature 500:477–481
3. Houseman EA, Accomando WP, Koestler DC et al (2012) DNA methylation arrays as surrogate measures of cell mixture distribution. BMC Bioinformatics 13:86
4. Emmert-Buck MR, Bonner RF, Smith PD et al (1996) Laser capture microdissection. Science 274:998–1001
5. Datta S, Malhotra L, Dickerson R et al (2015) Laser capture microdissection: big data from small samples. Histol Histopathol 30:1255–1269
6. Frommer M, McDonald LE, Millar DS et al (1992) A genomic sequencing protocol that yields a positive display of 5-methylcytosine residues in individual DNA strands. Proc Natl Acad Sci U S A 89:1827–1831
7. Susan JCi, Harrison J, Paul CL, Frommer M (1994) High sensitivity mapping of methylated cytosines. Nucleic Acids Res 22:2990–2997
8. Grunau C (2001) Bisulfite genomic sequencing: systematic investigation of critical experimental parameters. Nucleic Acids Res 29:65e–665
9. Olova N, Krueger F, Andrews S et al (2018) Comparison of whole-genome bisulfite sequencing library preparation strategies identifies sources of biases affecting DNA methylation data. Genome Biol 19:33
10. Smallwood SA, Lee HJ, Angermueller C et al (2014) Single-cell genome-wide bisulfite sequencing for assessing epigenetic heterogeneity. Nat Methods 11:817–820

11. Morrison J, Koeman JM, Johnson BK et al (2021) Evaluation of whole-genome DNA methylation sequencing library preparation protocols. Epigenetics Chromatin 14:28
12. Vaisvila R, Ponnaluri VKC, Sun Z et al (2021) Enzymatic methyl sequencing detects DNA methylation at single-base resolution from picograms of DNA. Genome Res 31:1280–1289
13. Han Y, Zheleznyakova GY, Marincevic-Zuniga Y et al (2022) Comparison of EM-seq and PBAT methylome library methods for low-input DNA. Epigenetics 17:1195–1204
14. Luo C, Hajkova P, Ecker JR (2018) Dynamic DNA methylation: in the right place at the right time. Science 361:1336–1340
15. Martisova A, Holcakova J, Izadi N et al (2021) DNA methylation in solid tumors: functions and methods of detection. Int J Mol Sci 22: 4247
16. Nishiyama A, Nakanishi M (2021) Navigating the DNA methylation landscape of cancer. Trends Genet 37:1012–1027
17. Lunnon K, Smith R, Hannon E et al (2014) Methylomic profiling implicates cortical deregulation of ANK1 in Alzheimer's disease. Nat Neurosci 17:1164–1170
18. Stroud H, Su SC, Hrvatin S et al (2017) Early-life gene expression in neurons modulates lasting epigenetic states. Cell 171:1151–1164.e16
19. Ziller MJ, Ortega JA, Quinlan KA et al (2018) Dissecting the functional consequences of De novo DNA methylation dynamics in human motor neuron differentiation and physiology. Cell Stem Cell 22:559–574.e9
20. Smith ZD, Meissner A (2013) DNA methylation: roles in mammalian development. Nat Rev Genet 14:204–220
21. Greenberg MVC, Bourc'his D (2019) The diverse roles of DNA methylation in mammalian development and disease. Nat Rev Mol Cell Biol 20:590–607
22. Baylin SB, Jones PA (2011) A decade of exploring the cancer epigenome — biological and translational implications. Nat Rev Cancer 11: 726–734
23. Conda-Forge Community (2015) The conda-forge Project: Community-based Software Distribution Built on the conda Package Format and Ecosystem
24. Martin M (2011) Cutadapt removes adapter sequences from high-throughput sequencing reads. EMBnetjournal 17:10
25. Krueger F, Andrews SR (2011) Bismark: a flexible aligner and methylation caller for bisulfite-Seq applications. Bioinforma Oxf Engl 27: 1571–1572
26. Langmead B, Salzberg SL (2012) Fast gapped-read alignment with bowtie 2. Nat Methods 9: 357–359
27. Danecek P, Bonfield JK, Liddle J et al (2021) Twelve years of SAMtools and BCFtools. Gigascience 10:giab008
28. Xue Y, Liu L, Zhang Y et al (2024) Unraveling the key role of chromatin structure in cancer development through epigenetic landscape characterization of oral cancer. Mol Cancer 23:190

Chapter 5

High-Resolution Laser Microdissection to Investigate Transcriptional States in Formalin-Fixed Paraffin-Embedded Samples

Pierluigi Di Chiaro, Lucia Nacci, and Giuseppe R. Diaferia

Abstract

The advent of spatially resolved transcriptomics has revolutionized our understanding of tissue heterogeneity. While advanced spatial transcriptomics (ST) technologies offer unparalleled and unbiased, whole-slide molecular mapping, laser microdissection (LMD) remains a highly valuable and often superior approach for specific research questions, particularly when coupled with RNA sequencing (RNA-seq). LMD provides high spatial resolution through precise morphological annotation, allowing for the isolation of defined cell clusters or even rare populations from complex tissues. This enables deeper and more sensitive transcriptomic profiling than is often achievable with some high-throughput ST methods that depend on computational deconvolution. Crucially, LMD is well established for use with routine histological samples, including challenging formalin-fixed paraffin-embedded (FFPE) specimens, facilitating invaluable retrospective studies on large patient cohorts. Unlike emerging ST platforms that can present significant per-sample costs and substantial computational overhead, targeted LMD-RNA-seq offers a more cost-effective and scalable solution for analyzing numerous specific regions across large numbers of patients. Therefore, a robust protocol outlining the use of LMD for routine histological samples, coupled with RNA-seq, remains an invaluable tool for discovery, enabling highly specific molecular insights from clinically relevant biospecimens.

Key words Laser microdissection, FFPE, Pancreatic cancer, RNA-seq, Spatial transcriptomics, Bioinformatics, Omics

1 Introduction

Laser microdissection (LMD) is a powerful technique that enables the precise isolation of selected cells or tissue regions directly from histological sections under microscopic visualization. It facilitates the collection of pure cell populations from complex tissue samples while preserving their spatial and morphological context, making it ideal for downstream molecular profiling [1]. This level of precision allows researchers to dissect the molecular contributions of distinct

Roberta Noberini (ed.), *Laser Capture Microdissection: Methods and Protocols*, Methods in Molecular Biology, vol. 3015, https://doi.org/10.1007/978-1-0716-5154-4_5,

cell types to both normal physiology and disease progression in their original context.

When LMD is coupled to RNA sequencing (RNA-seq), it represents an ideal combination to overcome some limitations observed with standard transcriptomic profiling techniques. Bulk RNA-seq, while widely used, averages transcriptional signals across mixed cell populations, thereby masking programs specific to rare or spatially distinct tumor compartments. On the other hand, single-cell RNA-seq (scRNA-seq) offers high-resolution profiling of individual cells but lacks spatial and morphological context, making it difficult to directly link transcriptional states to histological features [2, 3].

LMD bridges these gaps by enabling spatially resolved transcriptomics based on morphological annotation. A notable example is Geographical Position Sequencing (GEO-seq), which integrates LMD of relatively large numbers of cells from fresh or frozen tissue with RNA-seq, providing transcriptomic profiles across defined tissue areas. While GEO-seq enhances spatial resolution, it comes with key limitations: It requires high-quality RNA and optimally preserved tissues, is technically demanding, yields limited transcript coverage, and is incompatible with formalin-fixed paraffin-embedded (FFPE) samples, thus restricting its use in retrospective clinical studies [4]. In contrast, more recent and advanced spatial transcriptomic (ST) approaches, while offering unparalleled capabilities for unbiased discovery of spatial relationships within individual samples, face significant limitations. These include compromised true single-cell sensitivity, persistent challenges with reliable cell segmentation across diverse cell types, the generation of massive datasets requiring substantial computational infrastructure, and inherent low throughput that restricts the number of samples analyzed per study [5–7]. Consequently, LMD-RNA-seq remains a highly competitive, often superior, and dramatically more cost-effective approach for studying rare cell populations across a large number of patients or regions.

To overcome these limitations and extend spatial transcriptomics to more practical, clinical contexts, we developed a robust and reproducible workflow, LMD-seq, that minimizes contamination from neighboring cell types and preserves spatial fidelity. This method enables transcriptomic profiling of morphologically distinct small regions (approximately 200 cells) directly from formalin-fixed paraffin-embedded (FFPE) sections [8].

In this protocol, we provide a step-by-step guide to performing LMD-seq, including tissue preparation, microdissection of ~200 histologically defined cells, cell lysis, library preparation, and computational analysis tailored to low-input, fragmented RNA. We also introduce a reproducible bioinformatics framework optimized for short-read data derived from FFPE tissues, facilitating accurate expression quantification and downstream biological interpretation (*see* Fig. 1).

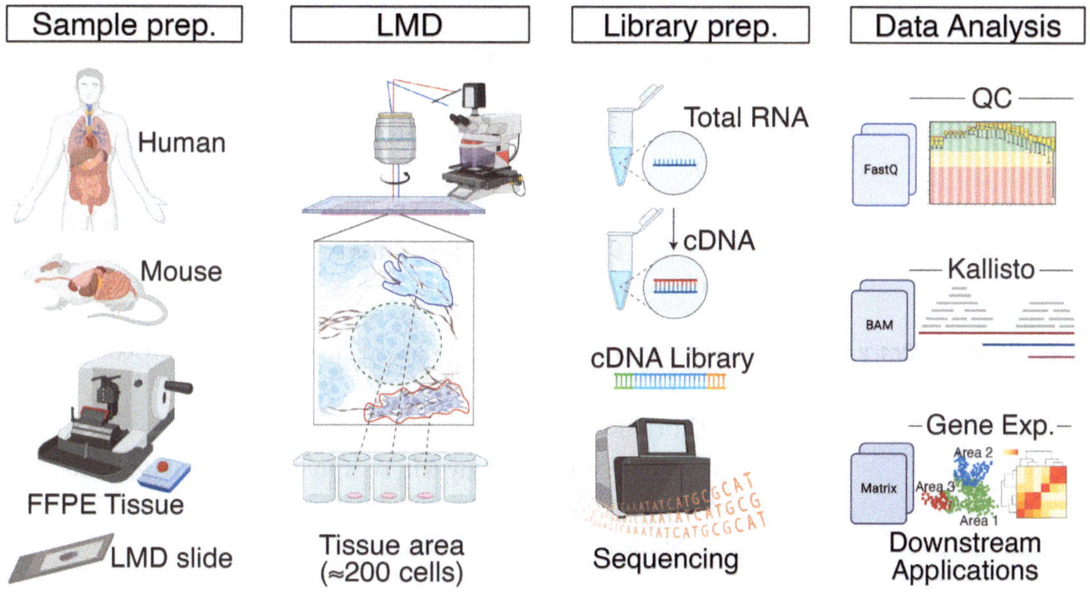

Fig. 1 Schematic workflow of the laser microdissection coupled with low-input RNA-seq (LMD-seq) approach to study transcriptional profiles of murine and human formalin-fixed paraffin-embedded (FFPE) tissue samples

2 Materials

2.1 Histology

1. RNAseZap® or an equivalent RNase decontaminant.
2. Polyethylene naphthalate (PEN) 4 μm membrane frame slides (Leica Microsystems).
3. UV crosslinker (254 nm wave light).
4. Microtome and water bath.
5. Microtome blades.
6. Nalgene Acrylic Desiccator Cabinet and silica gel beads.

2.2 Tissue Staining

1. Xylene, absolute ethanol, 85% ethanol, and 75% ethanol.
2. Cresyl Violet solution (1% Cresyl Violet in 75% EtOH).
3. 20 mM Tris–HCl, pH 8.0.
4. Falcon tube covered by aluminum foil.
5. Microscopy Supplies, EasyDip™ Slide Staining Jars.
6. Optical Microscope.
7. Vacuum chamber.
8. Chemical hood.

2.3 Laser Microdissection

1. UV-based LMD7 LMD System (Leica Microsystems).
2. Zerostat antistatic instrument (Sigma).

3. 0.2 mL 8-well strips or 0.2 mL tubes or 96-well PCR plates.
4. Thermal mixer with heated lid (Eppendorf ThermoMixer C© with ThermoTop©).
5. Digestion buffer (0.05% Triton X-100, 30 mM Tris–HCl pH 8.0).
6. 20 mg/mL Proteinase K (Thermo Fisher Scientific).
7. 5 mM Proteinase K inhibitor (dissolved in DMSO) (Calbiochem).

2.4 RNAseq Library

1. SMART Stranded RNA-Seq Kit (Takara, protocol version 051018).
2. AMPure beads.
3. Qubit dsDNA HS assay kit or Quant-iT PicoGreen dsDNA Assay kit.
4. TapeStation or BioAnalyzer.

2.5 Data Acquisition and Analysis

1. FastQC.
2. Trim Galore (v0.6.5).
3. Human genome (hg38); (https://emea.support.illumina.com/sequencing/sequencing_software/igenome.html).
4. GENCODE gene annotations (version 33); (https://ftp.ebi.ac.uk/pub/databases/gencode/Gencode_human/release_33/gencode.v33.basic.annotation.gtf.gz).
5. Chromosome size (hg38); (http://hgdownload.soe.ucsc.edu/goldenpath/hg38/bigZips/hg38.chrom.sizes).
6. Kallisto (v0.46.0).
7. IGV.
8. tximport R package (v1.14.2).
9. DESeq2 (v1.26.0).

3 Methods

3.1 Microtome Cutting and Mounting

Sectioning FFPE tissue for LMD demands careful handling to preserve RNA integrity and tissue morphology. PEN membrane slides facilitate precise laser cutting and efficient sample collection. While these steps largely follow standard histological protocols, strict RNase contamination control, using dedicated clean tools and thorough decontamination, is critical.

1. Photoactivate 4 μm PEN membrane slides (*see* **Note 1**) by placing them face up (membrane side up) in a short-wave UV light crosslinker for 30 min or under a UV lamp in a laminar flow hood for 1 h (*see* **Note 2**).

2. Apply RNaseZap® to a paper towel and wipe all exposed surfaces of the water bath and microtome. Repeat this step with 70% ethanol to rinse off any residues and let it dry.
3. Fill the water bath with molecular grade water and heat to 40–45 °C.
4. Chill paraffin-embedded tissue blocks on a cooling plate for 30 min before sectioning.
5. Use a new blade for each patient tissue block to cut 4 μm sections at the microtome. Using a clean brush, pick up the ribbon and briefly float it on ultrapure water at room temperature, then transfer it to a 40–45 °C water bath to fully flatten the section (*see* **Note 3**).
6. Cut at least three consecutive sections for LMD, hematoxylin and eosin (H&E) staining, and immunohistochemistry (IHC) for each paraffin-embedded tissue block. Ensure that the slides for LMD and IHC are consecutively oriented with the H&E-stained slide. This will allow to study the tissue morphology and identify areas for microdissection.
7. Mount sections onto the flat membrane side of the photoactivated LMD slides, avoid overlapping the tissue with the metal frame, and let them dry overnight at 37 °C (*see* **Note 4**).
8. Store the LMD slides in a dry place. For long-term storage, place them in a sealed, nuclease-free container with a desiccant to prevent moisture (e.g., a Nalgene Acrylic Desiccator Cabinet with silica gel beads).

3.2 Sample Preparation

1. Incubate the slides at 60 °C for ~ 30 min to improve tissue adhesion to the membrane and soften the paraffin.
2. Deparaffinize the sections by incubating them in xylene (two changes for 5 min each). Then rehydrate through a graded ethanol series: absolute ethanol (two changes for 3 min each), 85% ethanol for 1 minute, 75% ethanol for 1 min.
3. Stain the section for ~30 s with freshly prepared Cresyl Violet-buffered solution by mixing four parts of 1% Cresyl Violet solution with 1 part of 20 mM Tris–HCl, pH 8.0 (*see* **Note 5**).
4. Gently shake off the excess stain. Quickly rinse the section in 75% ethanol (at least 2–3 changes) to remove any residual stain. Check the staining quality under a microscope. If the staining is too weak, repeat **steps 3** and **4**.
5. Dry the sample completely in a vacuum chamber for 5 min. Alternatively, aspirate liquid in excess with vacuum or let it air-dry in a chemical hood for ~30 min or until fully dry.
6. Stained LMD slides can be immediately used for LMD or stored in a dry place with a desiccant to avoid any moisture, as described in **step 8** of Subheading 3.1.

3.3 Laser Microdissection

LMD is an efficient method for isolating cell clusters from fixed tissue sections under direct microscopic visualization. On the live microscopic image, specific regions of interest (e.g., individual cells or distinct tissue areas) are digitally outlined by the user. A focused UV laser then precisely cuts along the exact contour of these defined shapes, thereby isolating the desired tissue for collection. The isolated sample (or dissectate) falls by gravity into a collection vessel, such as standard thin-walled 96-well plates or 8–12 strip caps. These collectors are positioned directly beneath the sample, allowing the dissectates to fall straight into the tube caps by gravitational force [9].

In the following steps, we describe how to accurately and precisely microdissect a small number of cells (approximately 200) from a morphologically distinguishable tissue area. This precise dissection minimizes cross-contamination from surrounding cell types and preserves the integrity of transcriptomic profiles.

1. Turn on the LMD microscope at least 30 min before use to allow the system to equilibrate and the laser to warm up.
2. Use an anti-static device (e.g., ZeroStat 3) to discharge the tissue sections and the collection vessels (e.g., 8-cap strips), neutralizing any charged particles (*see* **Note 6**). Place the slides in the specimen holder on the scanning stage and the cap strip in the collector. The collector is located directly under the sample, allowing dissected cells to fall into the tube caps by gravity.
3. Manually identify and select the area of interest by observing the tissue under the microscope. Utilize information from the corresponding H&E-stained or IHC-immunostained consecutive sections to precisely identify the desired region or group of cells.
4. Adjust the following LMD cutting setting (*see* **Note 7**) to ensure precise and efficient cutting:
   - *Laser power*—Increase gradually until the tissue and membrane detach cleanly.
   - *Aperture*—Set to achieve a narrow, well-defined cutting line.
   - *Speed*—Use lower speeds to enhance cutting precision, especially in delicate or fragmented samples.
   - *Specimen balance*—Increase laser aperture at the end of the cut to ensure efficient detachment of the area.
5. Before cutting, calibrate the objective lens of interest following the instrument's user manual. Use a calibrated 10× objective for areas (*see* **Note 8**) larger than 10000 μm^2, or a 20× calibrated objective for smaller, more precise areas.

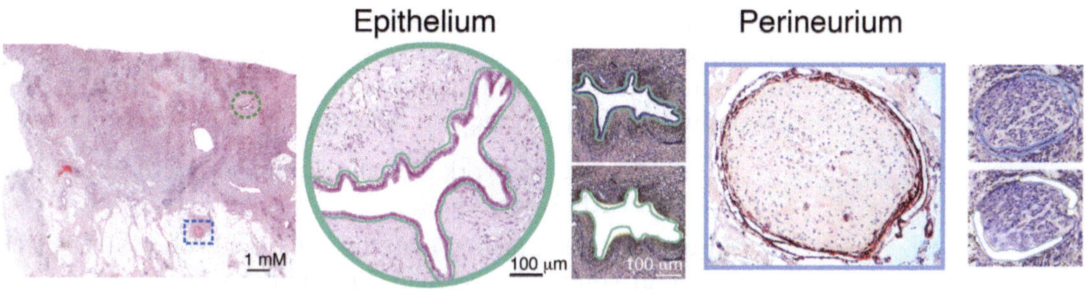

Fig. 2 Laser microdissection of tumor epithelium and perineurium regions in PDAC tissue. Hematoxylin and eosin (H&E)-stained images of a representative tumor epithelial region (red circle) and perineurium area surrounding a nerve fascicle (blue rectangle) from a PDAC patient. The corresponding regions were identified on a consecutive section stained with Cresyl Violet and used for laser microdissection (~200 cells per region). Images before (upper panels) and after (lower panels) microdissection are shown

6. Draw the region of interest and proceed with laser cutting, collecting the microdissected area into the cap of a collection tube. Capture images of the selected area both before and after cutting (*see* Fig. 2 and **Note 9**).
7. Inspect the collector for proper deposition of the isolated area into each cap and take a picture for documentation.
8. Before proceeding to sample digestion, export the object list with area measurements and then retrieve the strip caps and the LMD slide from the microscope stage (*see* **Note 10**).
9. Apply 5 μL of the digestion buffer supplemented with 1 μL of prediluted Proteinase K (10 U/mL) into the caps containing the microdissected samples (*see* **Note 11**). Place the strip caps onto the strip tubes, vortex briefly, and then spin down to place the material at the bottom of the tube (*see* **Note 12**).
10. Place the samples in a thermal mixer at 60 °C for 1 h, with mixing at 1000 rpm (*see* **Note 13**).
11. Chill samples on ice and spin down to collect any condensation at the bottom of the tubes.
12. Add 1 μL of 5 mM Proteinase K inhibitor to inactivate the digestion reaction.
13. Proceed immediately to RNA-seq library preparation or safely store samples at −80 °C until ready.

3.4 Library Preparation and Sequencing

This step enables the generation of indexed cDNA libraries directly from FFPE microdissected tissues and is adapted from the SMART-Seq Stranded RNA-Seq Kit (Takara). FFPE-derived laser microdissected samples, enzymatically digested using proteinase K and chemically inhibited [10], are used directly for library preparation without the need for RNA extraction to avoid sample loss.

In the following steps, we describe how library preparation and subsequent sequencing are performed according to the manufacturer's instructions, with minor modifications to accommodate the low-input, partially fragmented RNA typical of FFPE specimens.

1. Decontaminate the work area and equipment with RNaseZAP to prevent RNA degradation.
2. Ensure that your starting material is in a final volume of 7 μL. If the input volume is less than 7 μL, adjust it using RNase-free water. Keep all samples on ice.
3. Follow Option 2 (without fragmentation) of the manufacturer's instructions for the FFPE digested samples as the RNA is already partially fragmented due to fixation and embedding. It is recommended to include the provided positive control (purified total RNA from human brain) and to process it according to Option 1 (with fragmentation).
4. Follow the manufacturer's instructions for the subsequent main steps:
   - Perform the reverse transcription (RT) reaction for first-strand cDNA synthesis.
   - Carry out PCR-1 to amplify the cDNA.
   - Purify the amplified RNA-seq library using AMPure beads (*see* **Note 14**).
   - Deplete ribosomal cDNA using scZapR and scR-Probes.
   - Perform PCR-2 for RNA-Seq library amplification.
   - Quantify the resulting library using the Qubit dsDNA HS assay kit, Quant-iT PicoGreen dsDNA Assay kit, or equivalent method (Fig. 3).
   - Assess the library purity and size distribution using a TapeStation, BioAnalyzer, or equivalent instrument (Fig. 3).
5. Proceed to sequence the resulting sequencing libraries on an Illumina NovaSeq 6000 platform or equivalent instrument to obtain 50 bp paired-end reads. For sequencing libraries produced with this kit, Read 1 generates sequences antisense to the original RNA, reflecting the strand specificity of the SMART-Seq Stranded protocol (*see* **Note 15**).

3.5 Data Analysis

Transcriptomic analysis of FFPE-derived samples presents unique challenges due to RNA fragmentation and chemical modifications introduced during fixation and embedding. To address these limitations, we selected Kallisto [11], a pseudo-alignment-based method, as our primary tool for transcript quantification. Its speed, efficiency, and robustness against degraded RNA make it especially well suited for FFPE samples and low-input material such as those obtained via LMD.

A

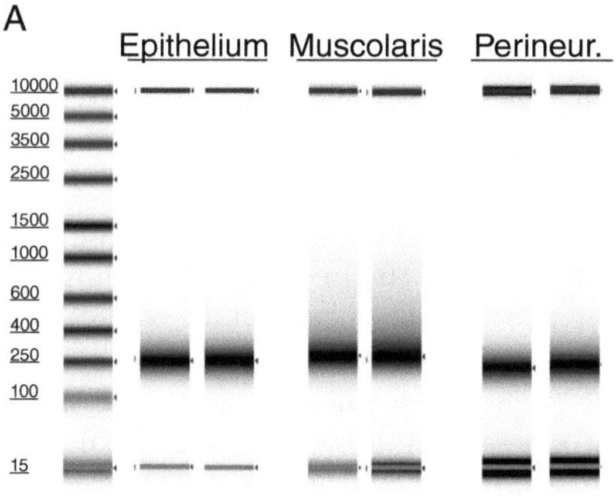

B

Sample	Cell Type	Yield (ng/μL)	Area (μm^2)
#1	Epithelium	6.4	24615
#2	Epithelium	4.6	24230
#3	Muscolaris	2.9	19848
#4	Muscolaris	2.7	13688
#5	Perineurium	2.7	9277
#6	Perineurium	1.3	7801

yield = ng/μL in 20 μL

Fig. 3 Quality and yield of RNA-seq libraries from microdissected FFPE hPDAC samples. (**a**) Profiles of RNAseq libraries generated from different FFPE tissue samples. (**b**) Table summarizing library yield and the microdissected tissue area for each sample

In this section, we show the most important steps of our computational workflow with the modifications and optimizations necessary to quantify RNAs derived from microdissected samples.

1. Perform FASTQ raw file preprocessing:
   - Run Trim Galore (v0.6.5) to filter out paired-end reads below a certain quality threshold and with high adapter contamination. Select the parameters, --clip_R2 3, --illumina, --stringency 3, and --length 20 (*see* **Note 16**).
   - Use FastQC to evaluate the quality trimming and adapter clipping. Go through the FastQC reports and check the sequence quality, read length, and adapter content accuracy.
2. Use Kallisto (v0.46.0) [11], a fast and memory-efficient tool, to employ the pseudo-alignment method to infer which transcripts are compatible with each read and use that information for accurate transcript quantification (*see* **Note 17**).
   - Build an appropriate index format to efficiently map the reads to a reference by running the *kallisto index* command selecting the *k-mer* length to 31 (*see* **Note 18**).
   - Run the pseudo-alignment and quantification algorithm on the human genome (GRCh38/hg38) using the *kallisto quant* command. *Kallisto quant* parameters: -b 50, --single-overhang, --rf-stranded, and –genomebam (*see* **Note 19**).
   - Check *run_info.json* files to verify the final number of unique sequencing reads (n_unique) (*see* **Note 20**) (Fig. 4).
   - Check the *abundance.h5* files in which transcript abundances are reported in "*estimated counts*" (est_counts).

- Visualize the transcript alignments to genome coordinates using a genome browser such as IGV (*see* **Note 21**).

3. Import abundance.h5 file data using tximport R package (v1.14.2) [12] to associate transcripts to gene IDs.
4. Keep only the protein-coding genes.
5. Filter out lowly expressed genes by retaining only those with sufficient read counts (at least 10) in at least two-thirds of samples within each patient and subsequently within each experimental condition or histological group.
6. Normalize raw read counts by a sample-specific size factor using the median of ratios method implemented in DESeq2 (v1.26.0) [13]. The size factors are derived to equalize the median of the ratios of gene counts to geometric mean per gene across all samples (Fig. 4).
7. Log_2-transform normalized counts and evaluate the distribution of read counts across samples (Fig. 4).

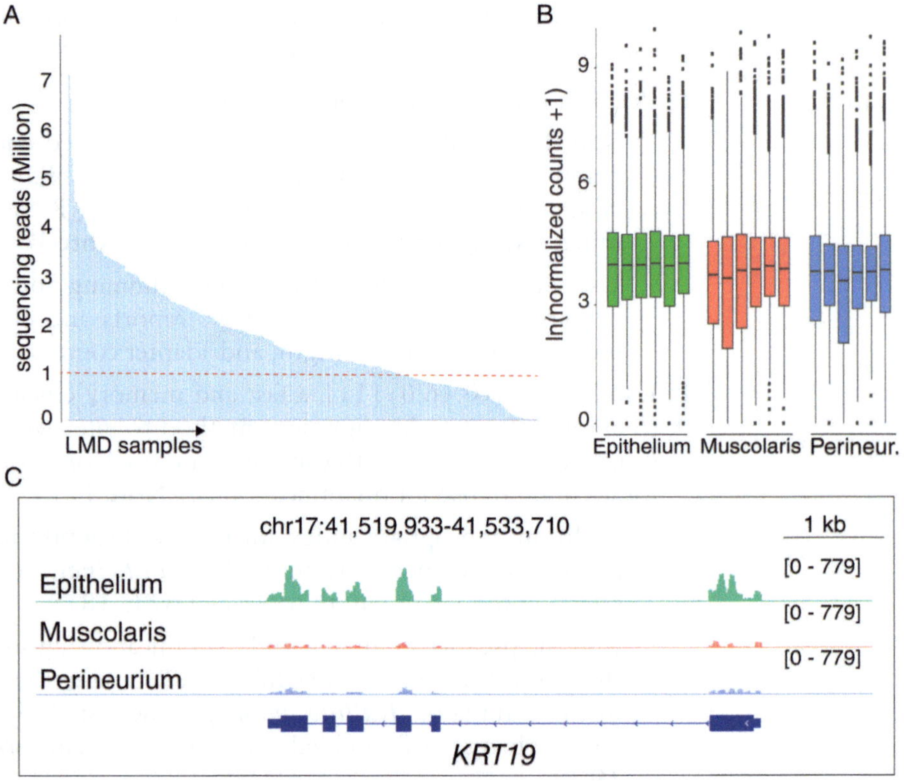

Fig. 4 Quality control and expression analysis of LMD-seq data. (**a**) Histogram showing the number of final sequencing reads for all microdissected samples. A horizontal red line indicates the minimum threshold used to retain high-quality samples. (**b**) Coverage distribution plots for representative microdissected samples, grouped by corresponding tissue region. (**c**) Representative RNA-seq coverage snapshot of a well-known epithelial marker across three microdissected tissue regions

8. Apply a variance stabilizing transformation (VST) using vst() function derived from the DESeq2 R package to the count data to stabilize variance across genes and reduce its dependence on the mean. Setting blind = TRUE allows for unbiased assessment of global expression patterns by not considering the experimental design.
9. Assess the overall data quality across all conditions by analyzing sample-to-sample distances and performing principal component analysis (PCA).
10. Perform differential expression analysis by running DESeq2 (v1.26.0), integrating both fold change and statistical significance to identify truly differentially expressed genes.
    - Conduct Wald test to compute *p*-values.
    - Adjust for multiple testing using the Benjamini–Hochberg method.
    - Define significant genes as those with adjusted p-value ≤ 0.01 and fold change ≥ 2 (either direction).
11. Use regularized $\log_2$-transformed (rlog) normalized counts for protein-coding genes to enable advanced downstream analyses such as clustering and subtype identification (*see* **Note 22**).

4 Notes

1. Several types of slides are available for LMD, each featuring specialized membranes designed to support high-quality tissue sectioning and dissection (https://www.leica-microsystems.com/science-lab/life-science/consumables-for-laser-microdissection). These include:
    - Glass Slides: These slides have a membrane—typically PEN (polyethylene naphthalate) or PPS (polyphenylene sulfide)—mounted on the upper surface and sealed at the edges. This configuration creates a rectangular mounting area with an air gap between the membrane and the glass. The membrane is specially coated to enhance specimen adhesion.
    - Frame Slides: These consist of a metal frame covered with one of several membrane options, including PEN, PPS, PET (polyethylene terephthalate), POL (polyester), or FLUO (fluorocarbon). In this format, tissue sections are mounted directly onto the membrane, which is tensioned across the frame. The raised structure of the frame ensures stability and uniform contact, facilitating optimal mounting and section integrity.

PEN membranes are the most used and are available in two standard thicknesses: 2.0 μm and 4.0 μm. The 4.0 μm membrane offers greater durability, improved section adherence, and reduced risk of membrane damage during mounting. Additionally, its increased weight enhances collection efficiency via gravity during LMD.

2. This step is essential to inactivate contaminating nucleases and to enhance tissue adherence. The membrane slides should be irradiated with a UV dose between 3 and 9 J/cm^2 to optimize hydrophilicity and enhance tissue adherence. A standard UV crosslinker with an output of 5 mW/cm^2 will deliver approximately 9 J/cm^2 in a 30-minute period. To ensure precise dosing regardless of bulb efficiency, set the instrument to Energy Mode, select the desired dosage (e.g., 3–9 J/cm^2), and the instrument will automatically calculate the time required. This treatment should be performed immediately before use, or if prepared the day before, slides must be stored in a sealed, nuclease-free container.
3. Briefly float tissue sections on ultrapure water at room temperature to gently smooth out any folds, ensuring this step is quick to avoid overexpansion and potential tissue damage. To prevent cross-contamination, never float sections from more than one tissue block simultaneously on the water bath.
4. If the tissue sections are small, you can place two consecutive sections on the same side of the flat membrane. However, make sure to avoid trapping bubbles under the sections during transfer and mounting process onto the LMD slides. Bubbles that get trapped beneath the sections can interfere with proper flattening and adhesion, which may affect the quality of the slide.
5. Apply approximately 250 μL of Cresyl Violet-buffered solution per slide to ensure the entire tissue section is evenly covered. Alternatively, immerse the slides in a nuclease-free container with enough staining solution to cover all sections. Store the 1% Cresyl Violet solution in a tightly sealed container to prevent evaporation and protect it from light. Discard any remaining buffered solution after use.
6. The use of an antistatic device is crucial for preventing the isolated tissue from electrostatically adhering to the membrane slide, which would hinder its gravity-driven fall into the collection vessel and result in sample loss. If this does not resolve the issue, dip the section in 75% ethanol and allow it to dry completely.
7. Set the laser to the lowest effective "*power*" to avoid burning the tissue section and adjust the "*aperture*" to control the width of the laser beam. The cutting "*speed*" should be set in

accordance with the laser power to achieve optimal cutting efficiency, allowing for the dissection of multiple objects within a reasonable timeframe. Furthermore, adjust the "*specimens balance*" parameter to ensure the efficient detachment of the microdissected area from the rest of the tissue at the end of the cut, thereby allowing for collection by gravity. Use the following parameters optimized for human pancreatic adenocarcinoma FFPE 4 μm tissue sections as a starting point: "33" laser power, "1" aperture, "7" speed, and "3" specimen balance. These parameters must be optimized based on the type of LMD slides, thickness and type of tissue, and optical objective.

8. A dehydrated pancreatic epithelial cell occupies approximately 150 μm^2, which can be used to estimate the number of cells within a dissected area. This estimation should be adjusted based on the specific cell type being collected as cell size may vary significantly depending on morphology and tissue context.

9. This protocol has been optimized for collection into 0.2 mL strip caps; however, individual 0.2 mL tubes or 96-well PCR plates are also suitable. When using individual tubes, ensure they have flat caps made of optical-grade plastic to allow visual inspection. For 96-well PCR plates, verifying the presence of collected material in the wells by manual inspection can be challenging.

10. LMD-stained slides can be reused for further microdissection if stored in a nuclease-free container inside a desiccator to prevent moisture accumulation and sample degradation. Additionally, object shapes can be safely saved with three calibration points before retrieving the slide from the microscope stage. This allows the shapes to be imported for future reference on the same slide.

11. Retrieve from the manufacturer's certificate of analysis the unit activity (U/mL or U/mg) of Proteinase K and dilute the enzyme to 10 U/mL in 50 mM Tris-HCl, pH 8.0, 3 mM $CaCl_2$, 50% glycerol. Store the diluted solution at −20 °C.

12. Microdissected samples might get stuck to the cap after the addition of the digestion solution and spinning. If this happens, increase the centrifugation to 10,000 rpm for up to 10 min or manually dislodge the microdissected material by pipetting the digestion solution in and out of the cap.

13. Ensure caps are tightly sealed to prevent evaporation and sample loss. Give particular attention to tubes located at the edges of the strip or the multiwell plate as these are often more prone to evaporation.

14. Accurate pipetting is critical for effective size selection using AMPure beads. The volume of AMPure beads used determines the molecular weight cutoff, which is essential for excluding undesired fragments (e.g., adapter dimers or overly long fragments). Please note that residual ethanol from the wash steps can significantly impact yield and downstream performance. It is therefore important to ensure complete removal of ethanol before elution. The performance of diluted AMPure beads for size selection was evaluated across various bead-to-sample volume ratios. In our protocol, a 0.7× bead-to-sample ratio was used to selectively retain fragments above the desired size threshold.
15. LMD-derived FFPE samples typically yield partially degraded and low-complexity libraries, which may require a sequencing depth of approximately 30–40 million reads per sample to achieve optimal saturation. However, the exact depth needed can vary depending on several factors, including the tissue type, the number of captured cells, and the heterogeneity of cell populations. It is recommended to assess sequencing saturation to determine whether the library has been sequenced deeply enough to capture the majority of expressed transcripts. If saturation has not been reached, additional sequencing may be necessary to improve transcript detection. In the current version of the kit, sequencing saturation and library complexity can be evaluated using tools such as Picard, which provides metrics including duplication rates. In newer versions of the kit that incorporate unique molecular identifiers (UMIs), saturation can be assessed more precisely by distinguishing PCR duplicates from true biological molecules.
16. During preprocessing, it is important to remove the first three nucleotides from Read 2, which correspond to the G-overhang introduced by the template-switching oligo. Other parameters, such as trimming stringency and minimum read length, were optimized specifically for FFPE samples, taking into account that RNA from these samples is typically short and partially fragmented.
17. The selection of Kallisto and its pseudo-alignment-based approach was the result of extensive testing and optimization, including comparisons with several full-length aligners such as STAR and TopHat2 [14, 15]. We found that Kallisto was particularly robust when handling fragmented or low-quality RNA, such as that derived from FFPE samples. It provided more accurate transcript-level quantification for short reads and effectively avoided the misalignments often observed with degraded RNA when using traditional aligners like STAR [21]. However, it is important to note that Kallisto

does not generate read-level alignment data, which may be required for certain downstream analyses such as variant calling, where full aligners like STAR remain more appropriate [8].

18. The k-mer length in Kallisto is a crucial parameter that influences the accuracy and efficiency of transcript quantification. It defines the length of substrings extracted from sequencing reads to identify matching transcripts through pseudo-alignment. Shorter k-mers can increase sensitivity but may result in ambiguous mappings, while longer k-mers improve specificity at the risk of losing reads with sequencing errors or polymorphisms. In our analysis, we tested several odd-numbered k-mer lengths and found that the default maximum value of 31 consistently yielded the best results. This setting provided an optimal balance between precision and robustness and was therefore used in all subsequent quantification steps.
19. Kallisto's pseudo-alignment algorithm is designed to infer transcript abundances from strand-specific short reads. In this case, parameters were applied to handle library strand orientation (--rf-stranded) and to retain reads extending beyond transcript boundaries (--single-overhang), thereby accounting for RNA fragmentation. Additionally, 50 bootstrap iterations (-b 50) were performed to estimate technical variance in transcript abundance.
20. To ensure data quality, we filter out sequencing reads below a defined threshold. In our dataset, we observed that samples with fewer than 1 million final unique reads tend to cause issues in downstream steps, particularly during sample normalization. Several factors must be considered when determining this threshold, including differences in cell types, sequencing depth, and experimental variability. Setting a minimum read count ensures robust comparisons and reliable normalization across all samples.
21. Kallisto constructs pseudo-alignments to the transcriptome but can project these alignments to genomic coordinates, enabling the visualization of split-read alignments in genome browsers such as IGV. However, similar to the scRNA-seq, it is generally not feasible or informative to visualize reads from individual low-input cells. Instead, it is recommended to aggregate reads from a specific cluster or cell group and visualize them collectively.
22. The versatility of the LMD-seq workflow, combined with its advanced downstream analysis capabilities, enables powerful applications across a wide range of biological and biomedical fields. It has been successfully employed for the reconstruction of gene regulatory networks [16], adapted for proteomics

[17], explored in the context of complex biophysical systems [18], leveraged to dissect intricate cell–cell communication dynamics [19], and integrated into cutting-edge AI-driven analytical pipelines [20]. These diverse applications underscore the impact of LMD-seq in both fundamental and translational research.

Acknowledgments

G.R.D was the recipient of the grant # GR-2016-02361721 by the Italian Ministry of Health and is supported by the Würth Foundation. P.D.C was partially supported by an AIRC fellowship for Italy, grant no. 25542.

References

1. Espina V, Wulfkuhle JD, Calvert VS et al (2006) Laser-capture microdissection. Nat Protoc 1(2):586–603. https://doi.org/10.1038/nprot.2006.85
2. Jovic D, Liang X, Zeng H et al (2022) Single-cell RNA sequencing technologies and applications: a brief overview. Clin Transl Med. 12(3): e694. https://doi.org/10.1002/ctm2.694
3. Haque A, Engel J, Teichmann SA et al (2017) A practical guide to single-cell RNA-sequencing for biomedical research and clinical applications. Genome Med. 9(1):75. https://doi.org/10.1186/s13073-017-0467-4
4. Chen J, Suo S, Tam P et al (2017) Spatial transcriptomic analysis of cryosectioned tissue samples with Geo-seq. Nat Protoc. 12:566–580. https://doi.org/10.1038/nprot.2017.003
5. Moses L, Pachter L (2022) Museum of spatial transcriptomics. Nat Methods. 19(5): 534–546. https://doi.org/10.1038/s41592-022-01409-2
6. Tian L, Chen F, Macosko EZ (2023) The expanding vistas of spatial transcriptomics. Nat Biotechnol. 41(6):773–782. https://doi.org/10.1038/s41587-022-01448-2
7. Carstens JL, Krishnan SN, Rao A et al (2024) Spatial multiplexing and omics. Nat Rev Methods Primers 4:54. https://doi.org/10.1038/s43586-024-00330-6
8. Di Chiaro P, Nacci L, Arco F et al (2024) Mapping functional to morphological variation reveals the basis of regional extracellular matrix subversion and nerve invasion in pancreatic cancer. Cancer Cell. 42 4:662–81e10. https://doi.org/10.1016/j.ccell.2024.02.017
9. Gross A, Schoendube J, Zimmermann S et al (2015) Technologies for single-cell isolation. Int J Mol Sci. 16(8):16897–16919. https://doi.org/10.3390/ijms160816897
10. Foley JW, Zhu C, Jolivet P et al (2019) Gene expression profiling of single cells from archival tissue with laser-capture microdissection and Smart-3SEQ. Genome Res. 29(11): 1816–1825. https://doi.org/10.1101/gr.234807.118
11. Bray NL, Pimentel H, Melsted P et al (2016) Near-optimal probabilistic RNA-seq quantification. Nat Biotechnol. 34(5):525–527. https://doi.org/10.1038/nbt.3519
12. Soneson C, Love MI, Robinson MD (2015) Differential analyses for RNA-seq: transcript-level estimates improve gene-level inferences. F1000Res. 4:1521. https://doi.org/10.12688/f1000research.7563.2
13. Love MI, Huber W, Anders S (2014) Moderated estimation of fold change and dispersion for RNA-seq data with DESeq2. Genome Biol. 15(12):550. https://doi.org/10.1186/s13059-014-0550-8
14. Dobin A, Davis CA, Schlesinger F et al (2013) STAR: ultrafast universal RNA-seq aligner. Bioinformatics. 29(1):15–21. https://doi.org/10.1093/bioinformatics/bts635
15. Kim D, Pertea G, Trapnell C et al (2013) TopHat2: accurate alignment of transcriptomes in the presence of insertions, deletions

and gene fusions. Genome Biol. 14(4):R36. https://doi.org/10.1186/gb-2013-14-4-r36

16. Alfarano G, Audano M, Di Chiaro P et al (2023) Interferon regulatory factor 1 (IRF1) controls the metabolic programmes of low-grade pancreatic cancer cells. Gut. 72(1): 109–128. https://doi.org/10.1136/gutjnl-2021-325811
17. Mulè P, Fernandez-Perez D, Amato S et al (2024) WNT oncogenic transcription requires MYC suppression of lysosomal activity and EPCAM stabilization in gastric tumors. Gastroenterology. 167(5):903–918. https://doi.org/10.1053/j.gastro.2024.06.029
18. Holuigue H, Nacci L, Di Chiaro P et al (2023) Native extracellular matrix probes to target patient- and tissue-specific cell-microenvironment interactions by force spectroscopy. Nanoscale. 15(37):15382–15395. https://doi.org/10.1039/d3nr01568h
19. Arora C, Matic M, Bisceglia L et al (2024) The landscape of cancer-rewired GPCR signaling axes. Cell Genom. 4(5):100557. https://doi.org/10.1016/j.xgen.2024.100557
20. Carli F, Di Chiaro P, Morelli M et al (2025) Learning and actioning general principles of cancer cell drug sensitivity. Nat Commun. 16(1):1654. https://doi.org/10.1038/s41467-025-56827-5
21. Di Chiaro P, Diaferia GR, Natoli G, Barozzi I. A framework to mine laser microdissection-based omics data and uncover regulators of pancreatic cancer heterogeneity. Gigascience. 2025 Jan 6;14:giaf101. https://doi.org/10.1093/gigascience/giaf101

Chapter 6

Transcriptomic Analysis from Cryosectioned Tissue Samples Using Laser Capture Microdissection-Based Geo-Seq

Xianfa Yang, Yingying Chen, Yun Yang, and Naihe Jing

Abstract

Geographical position sequencing (Geo-seq), combining transcriptome sequencing and laser capture microdissection, enables comprehensive learning of gene expression within the spatial context. Geo-seq is robust for in-depth gene expression analyses from tissue section at specific position coordinate across various tissue types. The abundance of specific transcripts and the whole transcriptomic architecture can be determined following region-specific laser capture microdissection, RNA extraction, and cDNA preparation. Generally, Geo-seq enables high-throughput profiling of the transcriptomic architecture encompassing over 8000 genes from a single sampled spot.

Key words Laser capture microdissection, RNA isolation, Reverse transcription, Transcriptomic analysis

1 Introduction

Laser capture microdissection (LCM) enables selective microscopic procurement of cells of interest from a heterogeneous tissue section [1, 2]. This ever-evolving technology toward enhancing sampling precision and minimizing the damage of the molecular and cellular composition has now largely expanded the applicability in the field of life sciences and medicine [3–6].

Transcriptome analyses have opened avenues for systematically uncovering the molecular events underlying both physiological and pathological conditions [7–9]. Nowadays, with the rapid development of single-cell transcriptomics, researchers can explore gene expression landscape at an unprecedented resolution, facilitating the dissection of cellular heterogeneity and the identification of rare cell populations [10]. However, single-cell technologies usually lead to the destruction of positional information, due to mechanical or enzymatic dissociation of tissue clumps for preparing

Roberta Noberini (ed.), *Laser Capture Microdissection: Methods and Protocols*, Methods in Molecular Biology, vol. 3015, https://doi.org/10.1007/978-1-0716-5154-4_6,

single-cell suspension. The ability of precise spatial sampling at specific coordinates within tissue sections has made the technology of LCM a compelling solution for addressing this challenge [11]. Typically, LCM includes microscopic visualization and searching for cells of interest and then excises them with a laser beam to precisely target and isolate individual cells or cell populations [12, 13].

Geo-seq is built primarily based on leveraging LCM and single-cell RNA-seq technology (Smart-seq2) [14], with notable optimizations for cell morphology preservation, RNA recovery, and cDNA acquisition to achieve robust gene expression profiling [15]. With continuing advancements in integrating with single-cell multimodal dataset [16, 17], the Geo-seq platform demonstrates enhanced scalability and superior capability in resolving complex biological questions. In this chapter, we will describe a step-by-step protocol for the most typical implementation of the Geo-seq protocol (Fig. 1).

2 Materials

Prepare all solutions using nuclease-free ultrapure water and reagents. Prepare and store all reagents at room temperature unless indicated otherwise. Follow proper waste disposal procedures when disposing waste reagents.

2.1 Laser Capture Microdissection System Setup

1. RNase-Zap.
2. 75%, 85%, 95%, 100% (vol/vol) Ethanol Buffer (absolute ethanol freshly resolved in nuclease-free water).
3. CellCut laser microdissection system (Molecular Machines & Industries).
4. PEN membrane slide (Molecular Machines & Industries).
5. Isolation cap (Molecular Machines & Industries).
6. Cover-glass slide.

2.2 Tissue Cryosection Preparation

1. PBS, pH 7.4.
2. Tissue freezing medium (Leica Microsystems).
3. 75%, 85%, 95%, and 100% (vol/vol) Ethanol Buffer (absolute ethanol freshly resolved in nuclease-free water).
4. Cryomold.
5. Leica CM1950 Cryostat (Leica Microsystems).

2.3 Laser Capture Microdissection and RNA Extraction

1. 1% cresyl violet acetate (1% (wt/vol) cresyl violet in 75% ethanol solution).

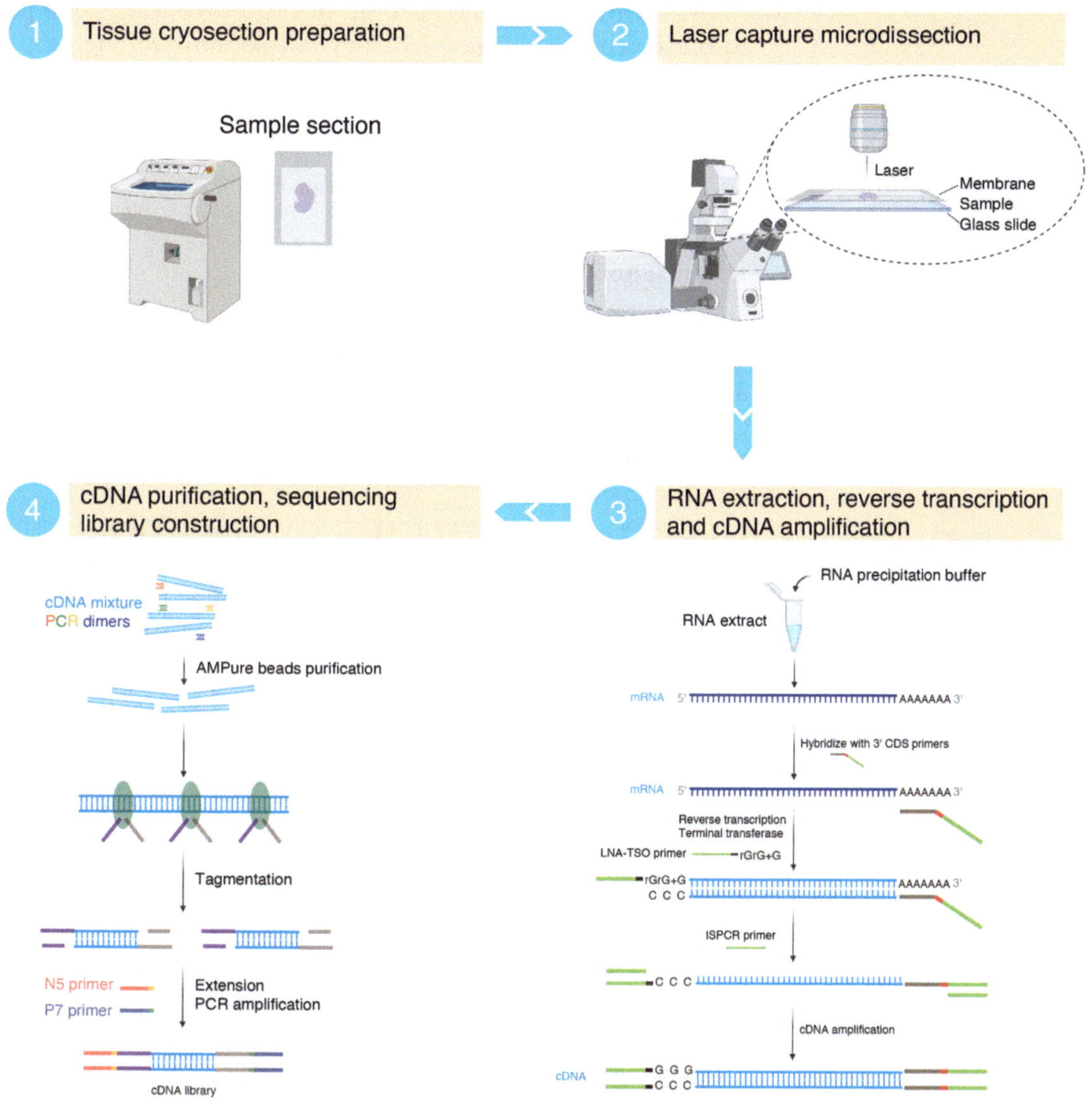

Fig. 1 The Geo-seq workflow

2. Guanidine isothiocyanate solution (4 M GuSCN, 50 mM Tris–HCL (pH 7.5), 25 mM EDTA).
3. 20 mg/mL glycogen.
4. 3 M pH 5.5 sodium acetate.
5. 75% and 100% Ethanol (vol/vol) (absolute ethanol freshly resolved in nuclease-free water).
6. Thermal incubator.
7. Microcentrifuge tube.

8. RNA precipitation buffer (150 μL nuclease-free water, 600 μL absolute ethanol, 20 μL 3 M pH 5.5 sodium acetate, 1 μL 20 mg/mL glycogen).

2.4 Reverse Transcription, cDNA Amplification

1. 2.5 mM each dNTP mix.
2. 40 U/μL Protector RNase Inhibitor.
3. 5 M Betaine solution.
4. 1 M magnesium chloride solution.
5. SuperScript II Reverse Transcriptase.
6. 3′ CDS primer (5′-AAGCAGTGGTATCAACGCAGAGTACT 30VN-3′, N represents any base; V represents A, C, or G. Dissolve the synthesized oligonucleotide powder using nuclease-free water to a final concentration of 100 μM; store at −20 °C before use).
7. LNA-TSO primer (5′-AAGCAGTGGTATCAACGCAGAGTA CATrGrG+G-3′, rG represents ribo-guanosine; +G represents LNA-modified guanosine. Dissolve synthesized oligos using nuclease-free water to a final concentration of 100 μM; store at −20 °C before use).
8. KAPA HiFi HotStart ReadyMix.
9. A-IS PCR primer (5′-AAGCAGTGGTATCAACGCAGAGT-3′, amine modification at the 5′ end with a C6 linker. Dissolve synthesized oligos using nuclease-free water to a final concentration of 100 μM; store at −20 °C before use).
10. Thermal cycler.
11. RNA dissolution buffer (3.41 μL nuclease-free water, 0.1 μL 100 μM 3' CDS primer, 1 μL 10 mM dNTP, 0.05 μL Protector RNase Inhibitor, 4.56 μL in total).
12. Reverse transcription mix (0.5 μL SuperScript II Reverse Transcriptase, 0.25 μL 40 U/μL Protector RNase Inhibitor, 2 μL 5× SuperScript II first-strand buffer, 0.5 μL 100 mM DTT, 2 μL 0.5 M Betaine, 0.09 μL 1 M magnesium chloride, 0.1 μL LNA-TSO primer, 5.44 μL in total).
13. Pre-amplification mix (12.5 μL KAPA HiFi HotStart ReadyMix, 0.1 μL 100 μM A-IS PCR primers, 2.4 μL nuclease-free water, 15 μL in total).

2.5 cDNA Purification, Quantitative Real-Time PCR, and Sequencing Library Preparation

1. VAHTS DNA Clean Beads.
2. 80% (vol/vol) ethanol solution.
3. Equalbit 1× dsDNA HS Assay Kit.
4. PCR tube.
5. TruePrep DNA Library Prep Kit V2 for Illumina.
6. ChamQ Universal SYBR qPCR Master Mix.

7. Magnetic rack.
8. Thermal cycler (Applied Biosystems).
9. 4200 TapeStation system (Agilent).
10. QuantStudio 3 Real-Time PCR Systems (Applied Biosystems).

3 Methods

Carry out all procedures at room temperature unless otherwise specified. All the reagents should be free of DNase and RNase, and experimental facilities should be pretreated with RNase-Zap. To ensure the performance, quality control is essential for critical experimental steps (Fig. 2).

3.1 Tissue Cryosection Preparation

1. Carefully collect samples (embryo, tumor, etc.) in cold PBS buffer. Carefully dissect the samples into appropriate sizes.
2. Gently transfer the sample into a cryomold and discard the surrounding PBS. Add pre-cold tissue freezing medium until the sample is completely embedded. Adjust the sample orientation in the mold as required.
3. Immediately freeze the tissue by placing the cryomold in liquid nitrogen or dry ice.
4. Setup the cryostat with chamber temperature at −22 °C and object temperature at −18 °C. Transfer the embedded sample into the cryostat chamber and affix to the metal stage. Allow the sample to equilibrate within the chamber for at least 30 min before sectioning.

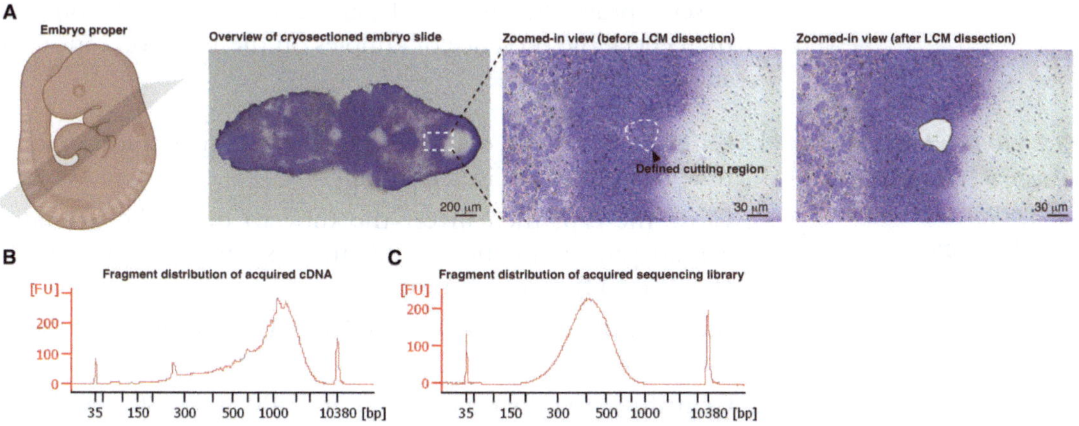

Fig. 2 Quality control of Geo-seq. (**a**) Representative images showing the laser microdissection of E9.5 mouse embryo cryosection slide sample. Fragment distribution analyses of acquired cDNA (**b**) and sequencing library (**c**)

5. Pretreat the PEN membrane slide and cover-glass slide with RNase-Zap, then wash with 75%, 85%, 95%, and 100% ethanol solution serially, and dry the slide at room temperature (*see* **Notes 1** and **2**).
6. Cryosection the samples at required thickness and carefully attach the tissue sections on the PEN membrane slide. Avoid placing the sample near the edge of the membrane slide (*see* **Note 3**).
7. Place the sample-attached membrane slide at room temperature for 15–30 min. Storing the slide in an RNase-Zap pretreated sample box with cover instead of open air is recommended to avoid potential contamination during this step (*see* **Notes 1** and **2**).

3.2 Laser Capture Microdissection

1. Carefully add ethanol solutions, with a serial order from 100%, 95%, 85%, to 75%, to the membrane slide and submerge the sample sections for 30 s.
2. Stain with 1% cresyl violet solution for 1 min. The cresyl violet treatment duration can be adjusted as required (*see* **Note 4**).
3. Dehydrate the post-stained sample slide with 75%, 85%, 95%, and 100% ethanol solutions and completely wash out the residual 1% cresyl violet.
4. Place the dehydrated slides at room temperature until dry.
5. Setup the CellCut system following the manufacturer's instructions. Calibrate the sample stage, laser focus, laser power, and cutting speed as required (*see* **Note 5**).
6. Place the sample slide above a cover-glass slide and then transfer it to the objective stage of the CellCut system. Find the region of interest with appropriate microscope objective lens.
7. Laser capture the target cells, and collect with isolation caps, then check the presence of samples on the Isolation caps using another microscope.

3.3 RNA Extraction, Reverse Transcription, and cDNA Amplification

1. Carefully add 50 μL of guanidine isothiocyanate solution to the tube bottom of the isolation caps.
2. Close the cap, then invert the tube to make the guanidine isothiocyanate solution, and fully submerge the samples attached to the isolation caps.
3. Incubate the tube at 42 °C for 15 min, vortexing the tube for every 5 min.
4. Centrifuge the collection tube at 7000 *g* for 30 s and then transfer the guanidine isothiocyanate lysis buffer to a new 1.5 mL microcentrifuge tube.

5. Add 771 μL of RNA precipitation buffer each sample. Mix thoroughly before incubation at −80 °C for at least 30 min.
6. Precipitate the RNA by centrifuging at 12,000 *g* for 30 min at 4 °C, wash the pellet with 75% ethanol, and centrifuge for another 10 min.
7. Discard the ethanol solution supernatant and air-dry the RNA pellet briefly.
8. Resolve the RNA pellet by repeatedly pipetting 4.56 μL of the RNA dissolution buffer. Avoid bubble forming during this step (*see* **Note 6**).
9. Transfer the RNA resolved solution to a new PCR tube and incubate at 72 °C for 3 min in a thermal cycler.
10. Briefly centrifuge the PCR tube and then add 5.44 μL of reverse transcription mix. Mix gently and proceed immediately to the reverse transcription step.
11. Incubate the tube in a thermal cycler with the reverse transcription program (42 °C for 90 min, 10 cycles of 50 °C 2 min and 42 °C 2 min, 70 °C for 15 min, 4 °C hold).
12. Briefly centrifuge the PCR tube and then add 15 μL of pre-amplification mix.
13. Incubate the tube in a thermal cycler with the cDNA pre-amplification program (98 °C for 3 min, 2–20 cycles of 98 °C 20 s, 67 °C 15 s, and 72 °C 6 min, 72 °C 5 min, 4 °C hold).

3.4 cDNA Purification, Sequencing Library Construction

1. Take out the DNA clean beads from the refrigerator and keep at room temperature for at least 30 min. Resuspend the beads by inverting thoroughly before use (*see* **Note 7**).
2. Add 25 μL of nuclease-free water to the cDNA pre-amplification PCR product. Mix thoroughly by pipetting up and down.
3. Add 40 μL of DNA clean beads to the PCR product. Mix thoroughly by pipetting 10 times.
4. Incubate the tube at room temperature for 10 min to bind DNA.
5. Place the tube on the magnetic rack for about 5 min until the supernatant is clear. Carefully discard the supernatant.
6. Keep the tube on the magnetic rack and wash the beads with 200 μL of freshly prepared 80% ethanol without disturbing the beads. Incubate at room temperature for 30 s and carefully discard the supernatant. Repeat twice.
7. Keep the tube on the magnetic rack and air-dry the beads for 5 min.

8. Remove the tube from the rack and add 17 μL of nuclease-free water to resuspend the beads. Incubate at room temperature for 2 min.
9. Place the tube back on the magnetic rack until the supernatant is clear (~5 min). Transfer the supernatant (~15 μL) to a new PCR tube.
10. Determine the concentration of cDNA using Equalbit 1× dsDNA HS Assay Kit with the manufacturer's instruction. The size distribution of purified cDNA can be checked using fragment analyzer like Agilent 4200 TapeStation system. The size of a high-quality purified cDNA pool should be larger than 1 kb with minimal short fragments (*see* **Note 8**).
11. For transcriptomic sequencing library preparation, take out 1 ng purified cDNA product, and then follow the instructions from TruePrep DNA Library Prep Kit V2 for Illumina.
12. For real-time PCR, the purified cDNA should be diluted in nuclease-free water 1:10 to 1:100 before use.

4 Notes

1. PEN membrane slides must undergo RNase decontamination prior to use to prevent RNA degradation during sample processing.
2. The experimental workspace should be regularly decontaminated using UV irradiation and RNase-Zap to maintain a RNase-free environment.
3. Avoid positioning the sample too close to the edges of the PEN membrane slide as this may hinder efficient laser cutting and increase the risk of collection failure due to misalignment with the isolation caps.
4. The incubation time for staining with 1% cresyl violet solution should be optimized for each sample type through preliminary testing to avoid overstaining.
5. The laser of the microdissection system must be calibrated before use to ensure precision and efficiency during the dissection process.
6. During the RNA pellet resuspension step, the pellet should be pipetted for at least five times per sample to ensure complete dissolution.
7. DNA clean beads used for cDNA purification should be incubated at room temperature for at least 30 min before use to achieve optimal performance.
8. When assessing cDNA fragment size, an additional purification step should be performed if dimers are observed.

Acknowledgments

This work was supported in part by the National Natural Science Foundation of China (32470866, 32130030, 31900454), the Major Project of Guangzhou National Laboratory (GZNL2023A02005), and the Union Project by Guangzhou National Laboratory and State Key Laboratory of Respiratory Disease, Guangzhou Medical University (GZNL2024B01007).

Conflict of Interests The authors declare no conflict of interests.

References

1. Emmert-Buck MR, Bonner RF, Smith PD et al (1996) Laser capture microdissection. Science 274(5289):998–1001
2. Espina V, Heiby M, Pierobon M et al (2007) Laser capture microdissection technology. Expert Rev Mol Diagn 7(5):647–657
3. Bhamidipati T, Sinha M, Sen CK et al (2022) Laser capture microdissection in the spatial analysis of epigenetic modifications in skin: a comprehensive review. Oxidative Med Cell Longev 1:4127238
4. Ellis P, Moore L, Sanders MA et al (2021) Reliable detection of somatic mutations in solid tissues by laser-capture microdissection and low-input DNA sequencing. Nat Protoc 16(2):841–871
5. Shen S, Li J, Huo S et al (2021) Parallel, high-quality proteomic and targeted metabolomic quantification using laser capture microdissected tissues. Anal Chem 93(25):8711–8718
6. Rao BH, Souček P, Hlaváč V (2022) Laser capture microdissection: a gear for pancreatic cancer research. Int J Mol Sci 23(23):14566
7. Jiang Z, Zhou X, Li R et al (2015) Whole transcriptome analysis with sequencing: methods, challenges and potential solutions. Cell Mol Life Sci 72(18):3425–3439
8. Mutz K-O, Heilkenbrinker A, Lönne M et al (2013) Transcriptome analysis using next-generation sequencing. Curr Opin Biotechnol 24(1):22–30
9. Yang X, Chen Y, Yang Y et al (2024) The molecular and cellular choreography of early mammalian lung development. Med Rev 4(3): 192–206
10. Yuan GC, Cai L, Elowitz M et al (2017) Challenges and emerging directions in single-cell analysis. Genome Biol 18(1):84
11. Datta S, Malhotra L, Dickerson R et al (2015) Laser capture microdissection: big data from small samples. Histol Histopathol 30(11): 1255–1269
12. Meier-Ruge W, Bielser W, Remy E et al (1976) The laser in the Lowry technique for microdissection of freeze-dried tissue slices. Histochem J 8(4):387–401
13. Ya D, Zhang Y, Cui Q et al (2023) Application of spatial transcriptome technologies to neurological diseases. Front Cell Dev Biol 11: 1142923
14. Picelli S, Bjorklund AK, Faridani OR et al (2013) Smart-seq2 for sensitive full-length transcriptome profiling in single cells. Nat Methods 10(11):1096–1098
15. Chen J, Suo S, Tam PP et al (2017) Spatial transcriptomic analysis of cryosectioned tissue samples with Geo-seq. Nat Protoc 12(3): 566–580
16. Wang R, Yang X, Chen J et al (2023) Time space and single-cell resolved tissue lineage trajectories and laterality of body plan at gastrulation. Nat Commun 14(1):5675
17. Yang X, Xie B, Shen P et al (2024) Spatiotemporal regulatory logics of mouse gastrulation. bioRxiv 2024.12.22.630012

Chapter 7

Immunofluorescence Imaging-Guided Laser Microdissection for Ultralow Input Spatial Tissue Proteomics

Sonja Fritzsche and Fabian Coscia

Abstract

Laser microdissection (LMD) enables the precise isolation of specific cells or regions of interest from tissue sections, guiding downstream molecular analyses with high spatial resolution. In this protocol, we describe an optimized protocol combining whole-slide immunofluorescence imaging, LMD, and low-input liquid chromatography (LC) mass spectrometry (MS)-based proteomics of FFPE tissue sections.

Key words Laser microdissection, Tissue proteomics, Immunofluorescence staining, Deep visual proteomics

1 Introduction

Laser microdissection, also known as laser capture microdissection (LCM), has been employed in life sciences since the 1990s for a wide range of applications, including DNA-, RNA-, and protein-based assays. It enables the precise isolation of subcellular structures, single cells, or regions of interest (ROIs) from tissue sections or cell cultures for downstream molecular profiling [1]. Over the years, various devices have been developed utilizing distinct collection principles, such as gravity-based collection, adhesive cap or thermoplastic film-based transfer, and pressure catapulting into microtubes or well plates. Noncontact systems are particularly advantageous because of their reduced risk of contamination, making them better suited for molecular analyses [2]. Notably, LMD has recently experienced a resurgence, particularly within the field of ultrasensitive mass spectrometry-based proteomics. Based on novel automated workflows [3], sample collection for proteome profiling has significantly advanced and paved the way for methods such as deep visual proteomics, combining targeted

Roberta Noberini (ed.), *Laser Capture Microdissection: Methods and Protocols*, Methods in Molecular Biology, vol. 3015,
https://doi.org/10.1007/978-1-0716-5154-4_7,

and untargeted spatial proteomics [4, 5]. DVP combines antibody-based imaging (IHC or IF), automated LMD, and ultrasensitive liquid chromatography (LC) mass spectrometry (MS) for cell type and spatially resolved proteome profiling. However, efficient and reproducible sample preparation protocols are required to achieve high image quality and proteome coverage from minute tissue amounts while minimizing tissue loss. This chapter provides a detailed protocol for immunofluorescence microscopy-guided laser microdissection coupled with low-input proteomics and was optimized for archival (FFPE) tissue.

2 Materials

2.1 Slide Preparation

1. PPS frame slides (Leica).
2. Poly-L-lysine should be reconstituted at 0.1% (w/v) in Milli-Q water and stored in aliquots at −20 °C as a 10× stock solution. For use, dilute to a 1× working concentration (0.01%, w/v).

2.2 Immunofluorescence Staining

1. NeoClear™ (Sigma-Aldrich).
2. Antigen retrieval buffer: EnVision FLEX Target Retrieval Solution High pH (50×) (Dako) should be diluted to a 1× working concentration in Mili-Q water for use.
3. Super PAP Pen Liquid Blocker (Science Services).
4. Blocking buffer: Prepare 3% aliquots of BSA (albumin bovine fraction V, protease-free, low IgG; Serva) in 1× PBS and store at −20 °C. Thaw fresh aliquot before each use.
5. Hoechst 33342 (Thermo Fisher). Dilute to 1 μg/mL working solution in 1× PBS freshly before each use.
6. ProLong™ Diamond Antifade Mountant (Invitrogen).

2.3 H&E Staining

1. Hematoxylin solution.
2. Eosin solution.
3. Aqua-Poly/Mount (Polysciences).

2.4 Laser Microdissection

1. QuPath software: http://qupath.github.io/.
2. 384-well low-binding plate (Eppendorf).
3. PCR ComfortLid (Hamilton).

2.5 Sample Preparation

Use LC-MS grade water for all dilutions during sample preparation.

1. Lysis buffer: 5 mM tris(2-carboxyethyl)phosphine hydrochloride (TCEP), 20 mM 2-chloroacetamide (CAA), 100 mM triethylammonium bicarbonate pH 8.5 (TEAB), 0.1% n-dodecyl-beta-maltoside (DDM) in water.

2. LysC digestion: Prepare 0.2 μg/μL (in water) LysC protease stocks. Before use, prepare 2 ng/μL in 0.1 M TEAB and 30% ACN in water.
3. Trypsin digestion: Prepare 0.1 μg/μL (in water) trypsin protease stocks. Before use, prepare 2 ng/μL containing 10% ACN and 0.1 M TEAB (pH 8.5) in water.

2.6 Sample Purification

1. EvoTip pure (Evosep).
2. Buffer A: 0.1% formic acid in LC-MS grade water.
3. Buffer B: 0.1% formic acid in acetonitrile.

3 Methods

3.1 Preparation of PPS Frame Slides for Laser Microdissection

To enhance tissue adherence to the slides, it is recommended to pre-coat the slides with poly-L-lysine. Poly-L-lysine forms a positively charged layer on the slide surface, which promotes electrostatic interactions with negatively charged components of formalin-fixed, paraffin-embedded (FFPE) tissue. This electrostatic attraction significantly improves tissue adhesion during mounting (*see* **Note 1**) (Fig. 1).

1. Submerge the slides in 70% ethanol for 5 minutes to remove surface contaminants.
2. Rinse the slides by dipping them in Milli-Q water three times to remove residual ethanol.
3. Allow the slides to air-dry completely at room temperature.
4. While the slides are drying, thaw the 10× poly-L-lysine (PLL) stock solution to room temperature. Dilute the stock in Milli-Q water to obtain a 1× working solution.
5. Once the slides are completely dry, apply the 1× poly-L-lysine (PLL) solution to the flat surface of the PPS frame slide with continuous membrane by pipetting directly onto the membrane. Incubate for 10 minutes at room temperature to ensure uniform coating.

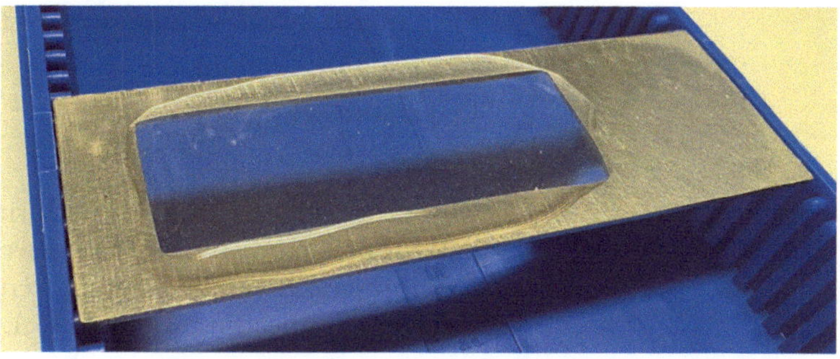

Fig. 1 Poly-L-lysine coating of PPS frame slides. PLL is applied to the frame side of the slide, which is entirely covered by the membrane

6. Carefully remove the PLL solution by gently tapping the slides to avoid uneven drying or damage.
7. Allow the coated slides to dry either at 60 °C for a minimum of 1 hour or at room temperature overnight.

3.2 FFPE Tissue Sectioning Using a Microtome

1. Fill the microtome water bath with Milli-Q water and heat it to 50 °C. This temperature facilitates even tissue flattening prior to slide mounting.
2. Place the cooling tube connected to the block holder in an ice bath to ensure consistent and optimal cooling of the tissue block during sectioning, which improves section quality and reduces artifact formation.
3. Additionally, FFPE tissue blocks should be briefly cooled at −20 °C for 5 minutes immediately prior to cutting (*see* **Note 2**).
4. Secure the tissue block in the block holder and adjust its orientation to ensure consistent and even sectioning.
5. Cut tissue sections at a thickness of 5–10 μm, depending on the tissue type. Allow the sections to float in the 50 °C water bath until fully flattened, which may take several minutes depending on fixation quality and tissue morphology (*see* **Notes 3** and **4**).
6. Mount the flattened sections by gently submerging the slide into the water and slowly lifting it from beneath the tissue at a 45° angle, allowing the section to adhere smoothly to the slide's surface. Take care to avoid air bubbles between the tissue and the slide.
7. Place the slides in a 37 °C incubator overnight to promote optimal tissue adherence.

3.3 Immunofluorescence Staining of FFPE Tissue

1. Incubate the slides at 60 °C for 30–60 minutes to enhance tissue adhesion and soften paraffin for efficient removal (*see* **Note 5**).
2. Submerge the preheated slides in NeoClear® for 5 minutes, then repeat in a fresh NeoClear® bath to ensure complete paraffin removal.
3. Gradually rehydrate the tissue sections by incubating the slides for 2 minutes each in a graded ethanol series:
   - 2× in 99% ethanol.
   - 1× in 80% ethanol
   - 1× in 70% ethanol.
4. Rinse the slides in 1× PBS, performing two consecutive incubations of 1 minute each to remove residual ethanol. If immunofluorescence staining is not required, proceed directly to **step 3** in Subheading 3.5 for H&E staining (*see* **Note 6**).

5. Perform heat-induced antigen retrieval by incubating the samples in a vertical jar filled with antigen retrieval buffer at 100 °C for 30 minutes using a steamer. Formalin fixation of FFPE tissue results in protein cross-linking, which can obscure antigenic epitopes and hinder antibody binding. Heat-mediated antigen retrieval restores epitope accessibility by reversing these cross-links. A retrieval buffer at pH 9.0 is effective for many antibodies; however, conditions may require optimization depending on the specific antigen and antibody used.
6. After antigen retrieval, remove the jar from the steamer and allow slides to cool to room temperature in the same buffer for 30 minutes.
7. Wash slides in 1× PBS three times for 5 minutes each to remove residual retrieval buffer.
8. To confine staining reagents and minimize usage, carefully draw a hydrophobic barrier around the tissue using a PAP pen. Ensure the line is placed at a safe distance from the tissue to prevent contact. Avoid touching any residual buffer with the pen ink as this may obscure the tissue and interfere with antibody binding.
9. Block nonspecific binding by incubating samples in blocking buffer for 30–60 minutes at room temperature.
10. While blocking, prepare primary antibody dilutions in blocking buffer at validated concentrations. Apply the antibody solution and incubate overnight at 4 °C in a humidified chamber.
11. The next day, wash slides in 1× PBS three times for 5 minutes each to remove unbound primary antibody.
12. Prepare the secondary antibody in blocking buffer at the appropriate dilution and incubate slides for 1–2 hours at room temperature in a humidified chamber, depending on antibody performance.
13. Gently tap off excess antibody solution and rinse slides by dipping several times in 1× PBS to remove unbound secondary antibody.
14. Counterstain nuclei by incubating slides in Hoechst solution for 10 minutes at room temperature.
15. Wash slides again in 1× PBS three times for 5 minutes each.
16. Mount coverslips using ProLong™ Diamond Antifade Mountant to preserve fluorescence and reduce photobleaching, then proceed with imaging. After imaging, promptly remove the coverslip to avoid adhesion that may hinder removal and damage the tissue (*see* Subheading 3.5, **step 1**). An example of immunofluorescence staining of a human FFPE tissue section is shown in Fig. 2a.

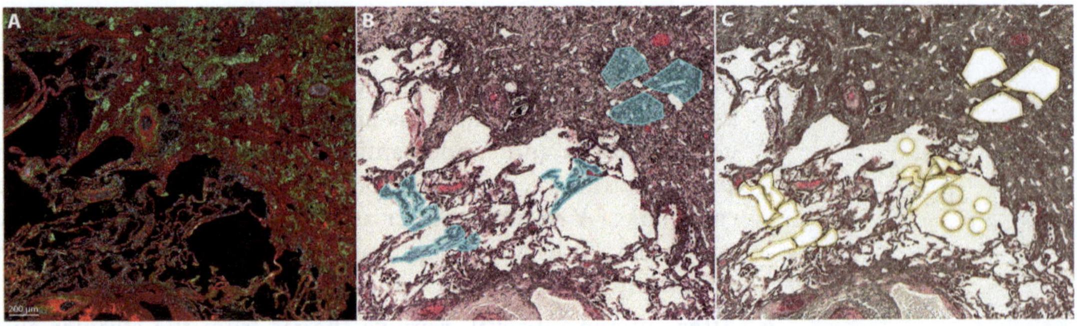

Fig. 2 Imaging-guided laser microdissection. (**a**) Immunofluorescence staining of a human FFPE tissue section (refer to Subheading 3.3). (**b**) H&E staining (*see* Subheading 3.4) with annotated regions of interest (ROIs) using QuPath (*see* Subheading 3.5). (c) Tissue section following laser microdissection (Subheading 3.6); black arrows indicate test cuttings

3.4 H&E Staining of FFPE Tissue

1. Immerse the slides in 1× PBS for 5 minutes to allow the coverslip to detach. Do not remove the coverslip manually as this may result in tissue loss.
2. Wash the slides in 1× PBS three times for 5 minutes each to remove any residual mounting medium.
3. Incubate the slides in hematoxylin solution for 3 minutes to stain cell nuclei. Hematoxylin is a basic dye that binds to acidic components such as DNA and RNA, resulting in a reddish to purple nuclear staining (*see* **Note 7**).
4. Rinse the slides in tap water for 10 minutes to shift the pH and stabilize the hematoxylin–DNA complex. During this step, the nuclear staining will turn blue which enhances the contrast.
5. Briefly dip the slides in Milli-Q water several times to remove excess staining solution.
6. Incubate the slides in eosin solution for 30 seconds to stain the cytoplasm and extracellular matrix components pink.
7. Briefly dip the slides in Milli-Q water several times to remove excess eosin.
8. Dehydrate the tissue sections by dipping the slides 10 times each in a graded ethanol series: 70%, 80%, and twice in 99% ethanol. Allow the slides to air-dry completely.
9. Mount coverslips using Aqua-Poly/Mount mounting medium and proceed to imaging.
10. After imaging, immerse the slides in 1× PBS for 5 minutes to allow the coverslip to detach. Do not attempt to remove the coverslip manually as this may damage the tissue. Wash the slides three times in Milli-Q water for 5 minutes each to remove any remaining mounting medium.

11. Allow the slides to air-dry completely. Slides can be stored in a dark, dry environment for several weeks prior to laser capture microdissection; however, prompt processing after staining is recommended to maintain tissue quality.

3.5 Select Regions of Interest for Laser Microdissection

1. Use the fluorescent or H&E image to identify regions of interest (ROIs) in QuPath (an open-source software for bio-image analysis) (Fig. 2b).
2. Employ QuPath's annotation tools to delineate the selected regions.
3. For this protocol, we recommended to collect a minimum total area of 8000 μm^2 per well from 5 μm thick tissue sections. For targeted cell-type analysis, annotate morphologically similar cells within the same collection well to enable deeper proteomic profiling. In general, more collected tissue results in greater proteome coverage. However, we recommend to not exceed 200,000 μm^2 per collection region to ensure optimal collection and sample preparation in 384-well plates.
4. After defining all ROIs, place three spatial reference points using the "Add Points" function to aid in alignment during laser microdissection. For larger sampling areas (e.g., whole tissue sections or TMAs), multiple sets of three reference points must be defined to ensure precise registration (*see* **Note 8**).
5. Export both the annotation and reference point data as GeoJSON files. Convert these files to the Leica LMD-compatible. xml format using the QuPath-to-LMD tool developed by the Coscia group. Detailed instructions for using this conversion tool are available on the Coscia Lab GitHub repository: https://github.com/CosciaLab/Qupath_to_LMD.

3.6 Laser Microdissection Using the Leica LMD 7 System

1. For sample collection and later on sample preparation, use low-binding 384-well plates.
2. Load the plate into the Leica LMD system by clicking the left "*Unload (change specimen)*" button and then click on the "*Specimen*" button in the top panel.
3. Insert the slide into the slide holder with the tissue side facing down, then click the right "*Unload (collection device)*" button to insert the collection device (*see* **Note 9**).
4. Acquire an overview scan to facilitate navigation. Click "*Specimen Overview*" → "*Create Specimen Overview.*" Manually move to the top-left corner of the area of interest and select "*Save top/left position.*" Then move to the bottom-right corner and click "*Save bottom/right position,*" followed by "*Scan.*" The resulting image will appear in both the "*Specimen Overview*" window and the slide holder display. You can navigate to specific locations by double-clicking on these images (*see* **Note 10**).

5. To load annotations, import the converted.xml file via "*File*" → "*Import Shapes.*" Select either all or a subset of shapes based on your intended sampling area. If using the annotations and tissue for the first time, do not select "*Use last stored reference points for calibration*" or "*Load reference points from file.*" A prompt will appear requesting calibration using the three reference points. Move the tissue so that the reference point you selected aligns with the calibration circle in the live display, then confirm. Repeat for all three reference points. Use the same magnification for loading annotations and for cutting (*see* **Note 11**).
6. Your loaded annotations will appear in the "Shape List." Double-click on any shape to navigate directly to it or assign collection wells as needed.
7. To calibrate the laser, click "*Laser*" in the main panel. Move to an area of the slide without tissue, then click "*Calibrate*" in the laser control settings, and follow the on-screen instructions.
8. Select the "*Middle Pulse*" cutting mode from the display panel. This setting cuts along the shape's outline and delivers a weak central pulse at the end to aid in dislodging the sample from the tissue.
9. Adjust laser cutting parameters using a sacrificial area of tissue. Optimal settings vary depending on tissue type and objective. Use test shapes similar in size and geometry to your actual annotations.

   Laser Settings Overview:

   Power: Laser beam intensity (indirect).

   Aperture: Width of the cutting line.

   Speed: Velocity of the laser movement.

   Middle Pulse Count: Number of central pulses per shape.

   Final Pulse: Intensity of the final pulse relative to the cutting power.

   Head Current: Direct laser output.

   Pulse Frequency: Pulses per second.

   Offset: Alignment between visual and laser focus
10. Select one or more shapes for cutting. Multiple selections will be collected sequentially (Fig. 2c).
11. Unload the slides and the collection plate.

3.7 Well Inspection After Collection

1. Centrifuge the 384-well plate at 1000× *g* for 5 min to collect all material at the bottom of the wells.

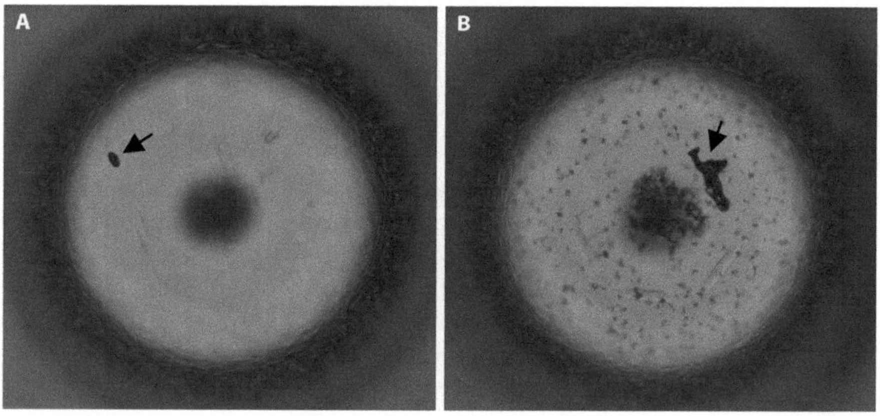

Fig. 3 Collected tissue in 384-well. (**a**) 3000 μm^2 and (**b**) 20,000 μm^2 samples collected at the bottom of the well

2. If samples remain on the well walls, rinse by gently pipetting 20 μL of acetonitrile (ACN) along the wall, then briefly centrifuge again.
3. Dry residual ACN using a SpeedVac at 60 °C for 30 min.
4. Reload the plate into the LMD system and select the "*Collection*" tab in the top panel.
5. In the "*Collection Device*" window (bottom left), select individual wells to inspect for successful collection. Ensure the well walls are also examined for remaining material (Fig. 3).
6. Samples are now ready for storage or further processing.

3.8 Sample Digestion for Liquid Chromatography Mass Spectrometry (LC-MS)

1. Prepare all buffers fresh before each step. Centrifuge the plate before and after each pipetting step to ensure that tissue and reagents are collected at the bottom of the wells (*see* **Note 12**).
2. Add 2 μL of freshly prepared lysis buffer to each well. Seal the plate with a ComfortLid and incubate at 95 °C for 1 hour in a thermal cycler with the lid set to 110 °C. Ensure the lid is tightly sealed to prevent evaporation, which could result in incomplete tissue lysis.
3. After the plate has cooled slightly, add 1 μL of Lys-C (4 ng/μL) to each well for protein digestion and incubate for 2–4 hours at 37 °C in a thermal cycler with the lid set to 50 °C. Avoid pipetting onto a hot plate as elevated temperatures can inactivate the enzyme.
4. Add 1 μL of trypsin (6 ng/μL) and incubate overnight at 37 °C in a thermal cycler. After this step, proteins will be fully digested into peptides.
5. Terminate the enzymatic reaction by drying the samples in a SpeedVac at 60 °C for 30 min (*see* **Note 13**).

3.9 Sample Purification for Liquid Chromatography Mass Spectrometry (LC-MS)

1. To remove salts and other soluble contaminants, clean the peptide samples using Evotips following the manufacturer's protocol. Prepare two containers filled with buffer A and buffer B to facilitate pipetting. Prepare an empty Evotip box containing buffer A at the bottom to keep the tips moist during handling.
2. Set up a box of empty Evotips, using one tip per sample, and place them in the Evotip box according to the sample layout. Prepare a second box with an equal number of tips as a counterweight for centrifugation. Fill a 96-well plate with 100 μL of isopropanol per well, matching the sample layout; this plate will be used to prime the Evotips.
3. To pre-wash the tips, add 20 μL of buffer B to each Evotip. Place both the sample box and the counterweight in the centrifuge and spin at 800× *g* for 60 seconds. Discard the flow-through.
4. To prime the tips, add 20 μL of buffer A to each Evotip and immediately transfer the tip rack into the isopropanol-filled 96-well plate. Soak for 10 seconds, ensuring that all tips appear uniformly pale white. If any tips are not pale, soak for an additional 10–20 seconds. From this step onward, it is important to always keep tips moist from the bottom by storing them in the buffer A-filled box during pipetting, and transfer to an empty box only for centrifugation.
5. Centrifuge the Evotips again at 800× *g* for 60 seconds and discard the flow-through. Immediately transfer the tips back into the buffer A-filled box.
6. Resuspend each sample in 10 μL of buffer A and transfer to its respective Evotip, allowing peptides to bind to the C18 material. Repeat the resuspension with 10 μL of buffer A and transfer from the same wells into the corresponding Evotips to maximize recovery. Centrifuge the tips at 800× *g* for 60 seconds and discard the flow-through.
7. Wash the Evotips by adding 20 μL of buffer A, followed by centrifugation at 800× *g* for 60 seconds. Discard the flow-through.
8. To maintain moisture, add 200 μL of buffer A and briefly centrifuge at 800× *g* for 10 seconds. The samples are now ready for LC-MS analysis and can be submitted directly from the Evotips using an Evosep device (*see* **Notes 14** and **15**). Begin mass spectrometry measurements as soon as possible after loading. Extended storage may result in peptide loss due to dissociation from the C18 matrix.

4 Notes

1. Slides may be coated in advance and stored for up to 1 week in a dry, dark environment at room temperature. However, for optimal tissue adhesion, it is recommended to freshly prepare the coated slides immediately prior to tissue sectioning.
2. For long-term preservation, blocks are best stored at room temperature in a dry, dark environment to maintain tissue integrity.
3. During sectioning, excess paraffin may accumulate on the blade. Periodically clean the blade using a soft brush to maintain cutting quality. To minimize the risk of injury, always engage the knife guard when the blade is not in use.
4. To optimize both staining quality and proteomic depth, thinner sections are recommended for tissues with high cellular density, while thicker sections are preferred for low cellular density.
5. It is not recommended to perform any steps on a platform shaker due to increased friction on the tissue, which can cause enhanced tissue loss.
6. To identify cells and regions of interest, immunofluorescence staining of several markers (depending on the hardware configuration of the microscope) can be performed prior to laser microdissection. For deep learning-based cell segmentation, it is highly recommended to stain for a nuclear marker (e.g., Hoechst) or a cell membrane marker.
7. Hematoxylin solution can be reused several times. Filter the solution using standard filter paper to remove any precipitates that may cause artifacts.
8. It is advisable to place reference points on the H&E image as this facilitates alignment with the Leica LMD7 microscope. If both H&E and fluorescence images are available, use the "Warpy" extension in QuPath to align them before annotation.
9. When the tissue is positioned face-down, its orientation will be mirrored relative to the previously captured image. This mirroring does not interfere with annotation alignment, which is handled by the software. However, it must be taken into account during reference point calibration.
10. Use a lower magnification for the overview scan to capture a larger area. If objectives are parfocal, this image can also be used to navigate under higher magnifications.
11. For optimal cutting efficiency, use the highest magnification that allows the entire annotation to be visible in the live window. Recommended guidelines:

63× for <6000 μm^2.

40× for <25,000 μm^2.

20× for <100,000 μm^2.

Additionally, round shapes cut more efficiently than angular ones; avoid sharp edges in your annotation.

12. For high-throughput workflows and improved pipetting accuracy, use an automated liquid handling system. Alternatively, manual pipetting can be performed using multichannel pipettes.
13. Samples can be stored at −20 °C or directly processed for peptide cleanup.
14. For optimal results of low-input tissue samples, samples should be analyzed on a high-sensitivity LC-MS setup, such as the timsTOF Ultra2 (Bruker) or Astral (Thermo Fisher) instruments connected to an Evosep ONE. The Whisper Zoom 40 SPD (samples per day) method facilitates good throughput and high sensitivity using the 15 cm Aurora Elite IonOpticks column. For the timsTOF Ultra2 system, MS acquisition should be performed using a standard dia-PASEF method (e.g., 3×8 window scheme) covering a m/z range of 400–1000 per scan.
15. Proteomic raw data analysis can be performed using common software tools, such as DIA-NN, MaxQuant, Spectronaut, or MSFragger.

Acknowledgments

The authors acknowledge funding support by the Federal Ministry of Education and Research (BMBF), as part of the National Research Initiatives for Mass Spectrometry in Systems Medicine, under grant agreement No. 161 L0222. This project received funding from the European Research Council (ERC) under the European Union's Horizon 2020 research and innovation program (grant agreement No. 101115681) and support by the ERC (ERC starting grant).

References

1. Emmert-Buck MR, Bonner RF, Smith PD et al (1996) Laser capture microdissection. Science (80-) 274:998–1001. https://doi.org/10.1126/science.274.5289.998
2. Datta S, Malhotra L, Dickerson R et al (2015) Laser capture microdissection: big data from small samples. Histol Histopathol 30:1255–1269. https://doi.org/10.14670/HH-11-622
3. Brasko C, Smith K, Molnar C et al (2018) Intelligent image-based in situ single-cell isolation. Nat Commun 9:1–7. https://doi.org/10.1038/s41467-017-02628-4
4. Mund A, Coscia F, Kriston A et al (2022) Deep visual proteomics defines single-cell identity and heterogeneity. Nat Biotechnol. https://doi.org/10.1038/s41587-022-01302-5
5. Horvath P, Coscia F (2025) Spatial proteomics in translational and clinical research. Mol Syst Biol:1–5. https://doi.org/10.1038/s44320-025-00101-9

Chapter 8

Low-Input Proteomics Protocols for Laser-Captured FFPE Samples with Orbitrap Astral DIA-MS

Daniele Musiani and Alessandro Cuomo

Abstract

Laser microdissection (LMD) combined with high-resolution mass spectrometry (MS) has become a powerful approach for analyzing targeted regions of fixed tissues, enabling detailed investigation of their heterogeneity. Here, we present two protocols optimized for low-input samples: a solid-phase enhanced preparation method (SP3) and a detergent/organic solvent-based approach (DDM/ACN). Both workflows are designed to minimize sample loss and are suitable for manual processing when automation platforms are not available. We provide practical guidance to reduce input requirements down to 1500 μm^2 of 10 μm thick FFPE tissue while maintaining reproducibility and proteome depth. These protocols, when coupled with fast instruments such as the Orbitrap Astral mass spectrometer working in DIA acquisition mode, enable sensitive, high-resolution proteomic analysis of microdissected tissue regions. Together, they establish a robust approach for investigating spatial heterogeneity in clinical samples with high coverage and precision.

Key words Laser microdissection, FFPE tissue, Low-input sample preparations for LC-MS/MS, Orbitrap Astral

1 Introduction

Laser microdissection (LMD) is a powerful tool for spatially resolved molecular profiling, allowing the precise isolation of defined regions from complex tissues. In recent years, LMD has increasingly been integrated into ultrasensitive mass spectrometry (MS)-based workflows, leading to significant advances in low-input spatial proteomics [1]. Despite this progress, sample preparation remains a key bottleneck—particularly with respect to efficiency and reproducibility—when working with very limited input material, such as microdissected tissue regions. These limitations are compounded by the limited availability of standardized protocols for manual processing in the absence of automation platforms. Among the available strategies, the sp3 and DDM/ACN-based methods have emerged as robust and scalable approaches for

Roberta Noberini (ed.), *Laser Capture Microdissection: Methods and Protocols*, Methods in Molecular Biology, vol. 3015, https://doi.org/10.1007/978-1-0716-5154-4_8,

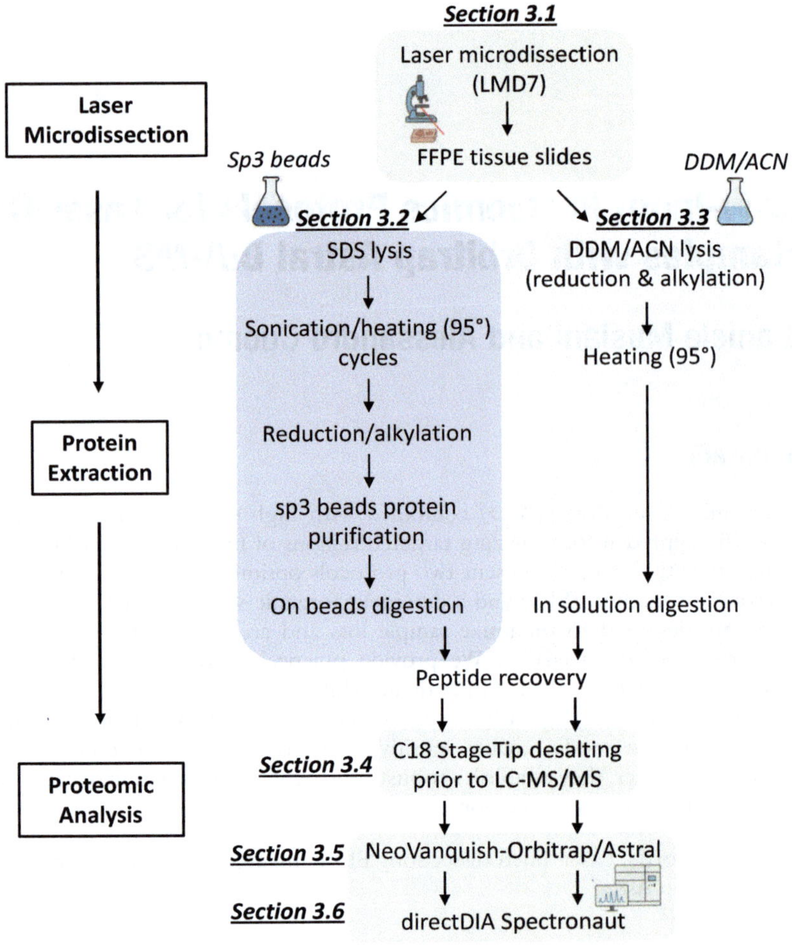

Fig. 1 Schematic overview of the sample preparation methods outlined in this chapter for LC-MS/MS-based proteomic analysis of low-input clinical samples. The cartoons were generated using ChatGPT (OpenAI)

peptide purification and digestion from clinical samples. Both workflows combine high recovery and reproducibility with minimal sample loss, making them particularly suitable for low-input material [2–4]. In this chapter, we focus on optimized protocols for LMD-derived FFPE tissue samples using sp3- and DDM/ACN-based preparation methods (Fig. 1). Recent advances in mass spectrometry, particularly the Orbitrap Astral, have revealed the performance of low-input dedicated workflows. When combined with Data-Independent Acquisition (DIA), the Astral mass spectrometer delivers high sensitivity and selectivity in a single analytical run, even for minute tissue samples. Furthermore, we evaluate complementary technologies such as Field Asymmetric Ion Mobility Spectrometry (FAIMS) to assess their contribution to peptide identification and proteome coverage in low-input conditions.

Combining efficient sample preparation and next-generation DIA-MS acquisition opens new opportunities for comprehensive proteome profiling of microdissected tissues, enabling the study of spatial heterogeneity with unprecedented depth, resolution, and reproducibility.

2 Materials

2.1 Laser Microdissection

1. Tissue sections, 10 μm thick and previously stained with hematoxylin/eosin, mounted on LMD slides.
2. 0.5 mL low-binding PCR tubes.
3. LMD7 Laser Microdissection system (Leica).
4. Leica Laser Microdissection Software.

2.2 Sample Preparation for LC-MS/MS Analysis

LC-MS grade water was used for all dilutions during sample preparation.

1. Acetonitrile for UHPLC-MS (ACN). Prepare an 80% ACN solution in water.
2. Ammonium bicarbonate (AMBIC). Prepare a 1 M stock solution, pH 8.0 in water, and store aliquots at −20 °C. Thaw stock aliquot and dilute as required in water freshly before each use.
3. 2-chloroacetamide (CAA): Prepare a 0.5 M stock solution in water and store at −20 °C, protected from light. Thaw aliquots in the dark as needed and dilute freshly before use in the appropriate buffer.
4. n-Dodecyl-beta-maltoside (DDM): Prepare 1% stock solution in water and store aliquots at −20 °C. Thaw stock aliquot and dilute as required in DDM Lysis Buffer freshly before each use.
5. Sera-Mag Speedbead carboxylate-modified [E3] (Cytiva) and Sera-Mag SpeedBead carboxylate-modified [E7] (Cytiva) magnetic particles. Store at 4 °C.
6. Trifluoroacetic acid (TFA): Store at room temperature in a glass container, protected from light. Always handle in a chemical fume hood.
7. C18 reversed-phase material (e.g., Empore C18 disks or equivalent).
8. 200 μL pipette tips.
9. Probe sonicator.
10. Bioruptor sonicator (Diagenode).
11. ThermoMixer.
12. Microcentrifuge.

13. SDS lysis buffer: 4% sodium dodecyl sulfate (SDS), 100 mM AMBIC, pH 8.0 in water.
14. 2× Reduction/Alkylation Buffer: 200 mM AMBIC pH 8.0, 20 mM tris(2-carboxyethyl)phosphine hydrochloride (TCEP), 80 mM CAA in water.
15. Digestion Buffer for sp3-based workflow: 50 mM AMBIC in water.
16. DDM Lysis Buffer for DDM/ACN-based workflow: 0.1% DDM, 5 mM TCEP, 20 mM CAA, and 100 mM triethylammonium bicarbonate pH 8.5 (TEAB).
17. Trypsin/LysC (20 μg lyophilized). Store at −20 °C. Prepare 0.5 μg/μL protease mix stock in storage buffer (provided with the proteases). Before use, prepare 10 ng/μL in either Digestion Buffer for sp3-based workflow or 10% ACN and 100 mM TEAB pH 8.5 in water for DDM/ACN-based workflow.
18. Buffer A: 0.1% formic acid (FA) in water (for LC/MS).
19. Buffer A*: 0.1% FA in 2% ACN. Prepare fresh as needed.
20. Buffer B: 0.1% FA in 80% ACN (for LC/MS).

2.3 LC-MS/MS Workflow for Sample Acquisition

1. NeoVanquish nanoHPLC system coupled to an Orbitrap Astral mass spectrometer equipped with Field Asymmetric Ion Mobility Spectrometry (FAIMS, Pro Duo) (Thermo Scientific).
2. Aurora Ultimate™ 25 cm x 75 μm ID, 1.7 μm C18 UHPLC column (IonOpticks).
3. Heater Controller (IonOpticks).
4. Xcalibur software (Thermo Scientific).

2.4 MS Data Analysis

1. HTRMS converter (Biognosys).
2. Spectronaut® 19 (Biognosys).

3 Methods

3.1 Laser Microdissection

The primary objective of this protocol is to guide readers through the sample processing and LC-MS/MS analysis of laser-microdissected FFPE tissue slides. Therefore, we have not covered slide preparation and staining as these steps have been thoroughly detailed in previous studies [2]. Nonetheless, here we shortly describe how we performed laser microdissection using the Leica LMD 7 system on breast cancer FFPE 10 μm thick tissue section.

1. Microscope magnification was set to 10x for selecting and dissecting the regions of interest (ROIs), while a 4x magnification was used for visual inspection of the 0.5 ml tube caps to confirm the presence of microdissected tissue.

2. Laser control parameters were configured as follows: Power at 42, Aperture at 5, and Speed at 5.
3. Using these parameters, we microdissected surface areas of FFPE breast cancer tissue sections in varying sizes: 750,000; 75,000; 37,500; 15,000; 7500; 3750 and 1500 μm^2.

3.2 sp3-Based Sample Preparation

We adapted the sp3-based protein purification and digestion protocol originally described in [5] to suit low-input samples derived from laser-microdissected FFPE tissues. Specifically, we successfully applied this modified protocol to laser-captured areas down to 75,000 μm^2, corresponding to approximately 150 cells [2] allowing the identification of more than 7000 protein groups per run (Fig. 2). It is recommended to reduce reagent volumes for input areas smaller than 75,000 μm^2 as outlined in Table 1. This protocol is optimized for use with samples collected in 0.5 ml Eppendorf tubes but can be easily adapted to a 96-well plate format, making it suitable for high-throughput applications (*see* **Note 1**).

1. Centrifuge the microdissected tissue containing tubes for 10 min, maximum speed at 4 °C. Make sure the microdissected tissues are at the bottom of the 0.5 ml tubes by visual inspection, when possible.
2. Resuspend the microdissected tissues in 10 μl of SDS lysis buffer (*see* **Note 2**).
3. Spin down briefly at maximum speed and sonicate for 25 cycles, 30′'ON/ 30′'OFF on HIGH at RT.
4. Incubate at 95 °C for 60 min (no shaking is needed). Repeat **steps 3** and **4** for a total of two times (*see* **Note 3**).
5. Quick spin down the microdissected tissue containing tubes to bring down droplets evaporated in the cap (15.000 g for 3 min).
6. Add 10 μL of 2× Reduction/Alkylation Buffer and incubate at 95 °C for 5 min.

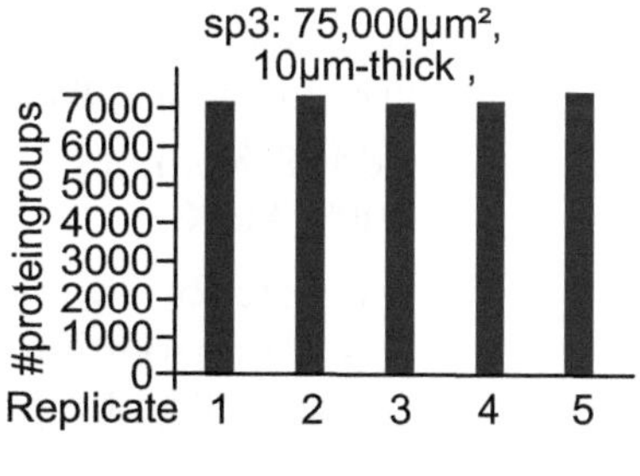

Fig. 2 Number of protein groups identified with Spectronaut v19 following sp3-based processing of laser-microdissected FFPE breast cancer tissue regions (75,000 μm^2, 10 μm thick) ($N = 5$)

Table 1
Reagent volumes for the sp3-based sample preparation of the indicated laser-microdissected FFPE areas

	Microdissected area (μm^2), 10 μm thick					
Volume (ml)	**75,000**	**37,500**	**15,000**	**7500**	**3750**	**1500**
SDS lysis buffer	10	10	5	5	3	3
2× reduction/alkylation (red-Alk) buffer	10	10	5	5	3	3
Final volume	25	25	15	15	11	11
100% ACN	75	75	45	45	33	33

7. Cool down at RT the microdissected tissue containing tubes (approximately 10–15 min) and meanwhile wash as described in point 8 the previously resuspended and equilibrated at RT sp3 beads.
8. Prepare a 1:1 mix of Sera-Mag Speedbead carboxylate-modified [E3] magnetic particles and Sera-Mag SpeedBead carboxylate-modified [E7] magnetic particles. Use 20 μg (10 μg Speedbead [E3] + 10 μg SpeedBead [E7]) of beads per sample and calculate the bead slurry volume to be taken according to the number of samples, with bead stock concentration being 50 μg/μL. Note that using less than 20 μg of beads per sample has proven difficult to handle reliably (*see* **Note 4**).
9. Wash the 1:1 bead mix with 1 mL of HPLC-grade water. Repeat this step for a total of three equilibrations. Place the tube on a magnetic separator and allow 2 min for the microparticles to clear.
10. After the last wash, resuspend the captured beads in HPLC-grade water to achieve a final concentration of 4 μg/μL. After gentle resuspension, add 5 μL of beads to each protein lysate equilibrated at RT (*see* **Note 5**).
11. Incubate the microdissected tissue + beads containing tubes at RT for 10 min while shaking at 1000 rpm. Double-check that possible aggregates are gone.
12. Add 3 volumes (25 μL × 3 = 75 μL) of HPLC-grade 100% ACN.
13. Incubate for 30 min at RT while shaking at 800 rpm. At this stage, the solution contains 75% ACN, the optimal solvent concentration for protein binding to sp3 beads.
14. Put the microdissected tissue + beads containing tubes on the magnetic rack for 2 min and discard the supernatant (use the pipette; do not use the vacuum).

15. Add 200 μL 80% ACN (do not pipette), slightly vortex (or flick the tubes with the finger) to resuspend the beads, and incubate at RT for 5 min while shaking at 1000 rpm.
16. Discard the supernatant and repeat the wash step as described in point 15.
17. After discarding the supernatant, add 200 μL 100% ACN (do not pipette), slightly vortex (or flick the tubes with the finger) to resuspend the beads, and incubate at RT for 5 min while shaking at 1000 rpm.
18. Discard the supernatant using the magnetic rack, dry the ACN for 15 min on the bench (cover with aluminum foil to avoid contamination), and add 25 μL of 50 mM AMBIC (Digestion Buffer).
19. Prepare a fresh Trypsin/LysC solution at a concentration of 20 ng/μL in Digestion Buffer and add 5 μL (100 ng) to the tubes containing laser-microdissected tissues and beads. Incubate the mixture overnight at 37 °C with shaking at 1000 rpm (*see* **Note 6**).
20. After 16 hours incubation, inactivate trypsin using 1% final TFA (0.8 μL) and incubate 5 min at RT while shaking at 1000 rpm (*see* **Note 7**).
21. After peptide recovery using the magnetic rack, directly transfer the eluate on C18 stage tips previously prepared and labeled as described in Subheading 3.4.
22. Perform a second peptide elution from the beads by adding 50 μl of HPLC-grade water. Incubate 5 min at RT while shaking at 1000 rpm and transfer the second elution onto the same C18 StageTips used for the first peptide elution.
23. Proceed to peptide desalting using C18 StageTips (Subheading 3.4) (*see* **Note 8**).

3.3 DDM/ACN-Based Sample Preparation

In this section, we describe the DDM/ACN-based sample preparation method [2] for LC-MS/MS analysis of laser-microdissected tissue sections (*see* **Note 9** and Fig. 3). In this section, we detail the DDM/ACN-based protocol as applied down to 1500 μm^2 microdissected areas from FFPE tissue samples (Table 2).

1. Centrifuge laser-microdissected tissue containing tubes for 10 min, at maximum speed at 4 °C. Make sure the laser-microdissected tissue are at the bottom of the 0.5 ml tubes by visual inspection.
2. Resuspend laser-microdissected tissues in DDM Lysis Buffer according to the volumes indicated in Table 2 "DDM Lysis Buffer, First Time" (*see* **Note 2**).
3. Spin down briefly at maximum speed and incubate at 95 °C for 60 min without shaking.

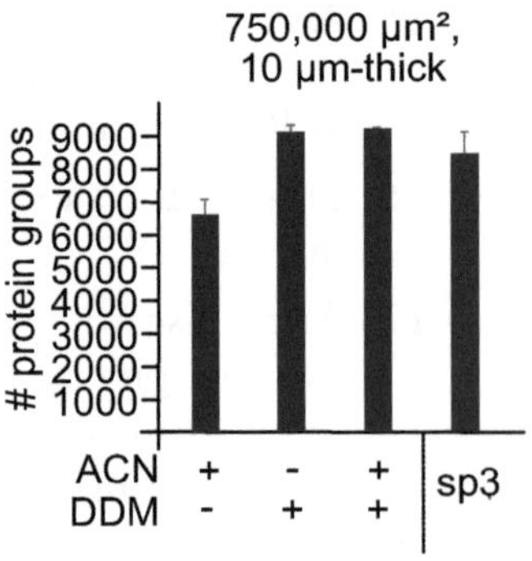

Fig. 3 Results of the comparative analysis of the methods described in this chapter upon processing of laser-microdissected FFPE breast cancer tissue areas (750,000 μm^2, 10 μm thick) ($N = 2$)

Table 2
Reagent volumes for the DDM/ACN-based sample preparation of the indicated laser-microdissected FFPE areas

	Microdissected area (μm^2), 10 μm thick					
Volume (mL)	**75,000**	**37,500**	**15,000**	**7500**	**3750**	**1500**
DDM lysis buffer, first time	5	5	3.5	3.5	2	2
DDM lysis buffer, second time	2.5	2.5	2.5	2.5	1	1
LysC/trypsin, 10 ng/ml	2.5	2.5	2	2	1	1
Buffer A	40	40	42	42	46	46
100% TFA	0.5	0.5	0.5	0.5	0.5	0.5

4. Spin down at 15.000 g for 3 min to bring down droplets evaporated in the cap and add DDM Lysis Buffer according to the volumes indicated in Table 1 "DDM Lysis Buffer, Second Time."
5. Cool down the samples and digest overnight at 37 °C without shaking with LysC/Trypsin solution according to the volumes indicated in Table 2 (10 ng/μL containing 10% ACN and 100 mM TEAB pH 8 in HPLC water).
6. After a 16-hour incubation, top up the samples to 50 μl with buffer A and then add 1% final TFA according to the volumes indicated in Table 2.
7. Incubate for 5 min at RT while shaking at 1000 rpm.
8. Recover the peptides and directly transfer the eluate on C18 StageTips previously prepared and labeled.
9. Proceed to peptide desalting using C18 stage tips (Subheading 3.4) (*see* **Note 8**).

3.4 C18 StageTip Peptide Desalting

This procedure purifies peptides prior to LC-MS/MS injection. Below, we describe the preparation of homemade C18 StageTips:

1. Assemble a micro-spin desalting column per sample by cutting two small disks (1–2 mm diameter) of C18 reversed-phase material from a C18 disk and inserting them into a 200 μL pipette tip.
2. Condition the StageTip by adding 80 μL 100% ACN and centrifuging at 1000 g for 3 min.
3. Add 80 μL of buffer B to the tip and centrifuge at 1000 g for 3 min. Repeat this step for a total of two times.
4. Equilibrate the tip by adding 80 μL buffer A and centrifuge at 1000 g for 3 min. Repeat this step for a total of two times.
5. Apply your peptide sample (prepared as described in Subheading 3.2 or 3.3) onto the equilibrated StageTip and spin at 500 × g for 5 min to allow peptides to bind. Ensure all sample passes through the plugs; if not, increase centrifuge speed.
6. Wash the StageTip with 80 μL buffer A and centrifuge at 1000 g for 3 min. Repeat this step for a total of two times.
7. Elute peptides by adding 80 μL buffer B and collect the eluate in a clean tube by centrifuging at 500 g for 3 min. Ensure all sample passes through the plugs; if not, increase centrifuge speed.
8. Dry down the eluted peptides in a vacuum centrifuge (SpeedVac).
9. Reconstitute peptides in 4 μL buffer A* for one-shot LC-MS/MS injection, except for the 5000-cell samples, which are resuspended in 10 μl buffer A* with 1 μl injected for LC-MS/MS.

3.5 LC-MS/MS Workflow for Sample Acquisition

Here, we describe the use of a NeoVanquish nanoHPLC system for peptide separation and subsequent injection into a Thermo Scientific Orbitrap Astral mass spectrometer equipped with FAIMS (Field Asymmetric Ion Mobility Spectrometry) Pro Duo when indicated (*see* **Note 10**).

1. To improve sensitivity and peptide identification in low-input samples (≤37,500 μm^2), FAIMS was used. In our hands, a fixed CV of −53 combined with a carrier gas flow of 3.5 provided the best performance; however, users should determine the optimal CV and carrier gas settings on their own instrument (*see* Fig. 4 and **Note 11**).
2. Assemble an IonOpticks Aurora Ultimate™ 25 cm × 75 μm ID, 1.7 μm C18 UHPLC column on a Thermo Scientific Easy-spray™ source. The column temperature was maintained at 50 °C using an IonOpticks Heater Controller (*see* **Note 12**).

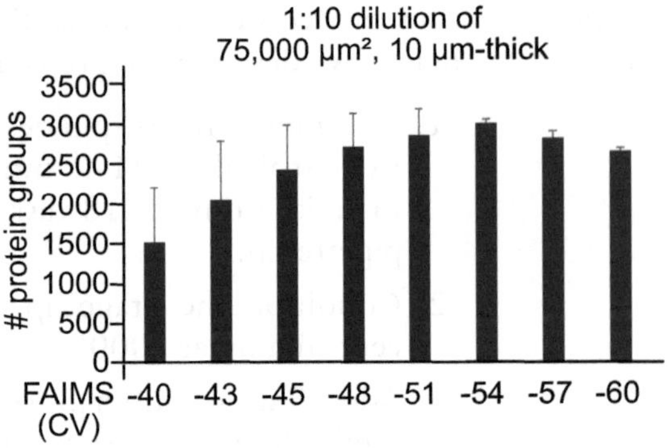

Fig. 4 Number of protein groups identified using Spectronaut 19 upon sp3-based processing of laser-microdissected FFPE breast cancer tissue areas (1:10 dilution of 75,000 μm^2, 10 μm thick) using the indicated FAIMS CV values ($N = 2$)

3. We recommend using a 25-minute chromatography method suitable for a 50 samples-per-day (50 SPD) throughput. This setup offers a practical balance between separation quality and analytical throughput. Peptide separation was performed using an initial gradient from 4% to 12% buffer B over 1.8 minutes at a flow rate of 450 nL/min, followed by a gradient from 12% to 22.5% over 10 minutes, and then from 22.5% to 40% buffer B over 7.5 minutes at 200 nL/min. This gradient was consistently applied across all experiments presented in the figures of this chapter.
4. For the 750,000 and 75,000 μm^2 microdissected areas, we used the MS method parameters summarized in Table 3. Nonetheless, to enhance sensitivity and improve peptide identification in low-input samples (i.e., microdissected areas ≤37,500 μm^2 from a 10 μm-thick slide), we installed the FAIMS (CV-53) and used an MS/MS isolation window of 5Th in combination with a MS/MS maximum injection time of 10 ms while keeping all other settings constant.

3.6 MS Data Analysis

1. Raw data are converted using the HTRMS converter and analyzed using Spectronaut® v19 (Biognosys), employing the directDIA™ library-free workflow (*see* **Notes 13** and **14**).
2. Searches are performed against the Homo Sapiens reference proteome (downloaded on the 5th of May 2025, Uniprot UP000005640, 83,587 entries) with trypsin specificity (up to two missed cleavages), fixed carbamidomethylation of cysteine, and variable oxidation of methionine and N-terminal acetylation. Minimum and maximum peptide lengths were 7 and 52 amino acids, respectively.

Table 3
MS method parameters used for the LC-MS/MS-based analysis of the clinical samples described in this chapter

Method duration (min) = 25
FAIMS mode = not installed
Experiment 1
Master scan: Full scan (MS)
Detector type = Orbitrap
Orbitrap resolution = 240,000
Scan range (m/z) = 380–980
Time (ms) = 30
Normalized AGC target (%) = 500
RF lens(%) = 40
EASY-IC = off
Experiment 2
Master scan: DIA
Isolation window (m/z) = 2
Window overlap (m/z) = 0
Detector type = astral
HCD collision energies (%) = 27
Precursor mass range (m/z) = 380–980
Scan range (m/z) = 150–2000
Maximum injection time (ms) = 3
Normalized AGC target (%) = 500
Polarity = positive
Time (sec) = 0.6

4 Notes

1. In our practice, we find that 0.5 mL tubes facilitate precise visual inspection of the microdissected tissues. By contrast, in the 96-well plate format, tissues often stick to the walls, making inspection more challenging.
2. We recommend gently opening the Eppendorf tube's cap just enough to insert the tip. In our hands, gently tapping the tube rather than pipetting minimizes sample loss.

3. In our workflow, we take the sp3 beads ([E3] and [E7]) out of the fridge about 30 minutes before use and keep them under mild agitation. This ensures the slurry reaches room temperature and becomes homogeneous before the wash steps.
4. We always prepare bead slurry in excess (enough for at least five additional samples) to account for pipetting variability and to avoid running short during processing.
5. We maintain the final bead concentration above 0.5 μg/μL in all cases, regardless of protein amount. This practice ensures consistent protein binding efficiency across different sample inputs.
6. To prepare the 20 ng/μL Trypsin/LysC solution, dilute a 500 ng/μL stock solution (previously prepared in storage buffer and stored at −20 °C for a maximum of 2 weeks) at a 1:25 ratio.
7. We calculate the 100% TFA volume (0.8 μL per sample) based on our recommendation to perform a second peptide elution with 50 μL of HPLC-grade water.
8. Both commercially available and homemade C18 StageTips can be used. However, for high-throughput workflows, we recommend streamlining peptide loading and separation by using Evotips (Evosep One, Evosep) directly coupled to the mass spectrometer according to manufacturer's instructions.
9. We compared the DDM/ACN-based sample preparation method to the sp3-based method described in the previous section. The results of this comparative LC-MS/MS analysis demonstrate that both methods are similarly efficient in purifying peptides from a relatively high input sample as 750,000 μm^2 (Fig. 3). In this comparative analysis, we also included the other two methods described in (2), namely, ACN-based and DDM-based. Our results indicate that the DDM + ACN- and sp3-based sample workflows outperformed the ACN-based method.
10. The use of Evosep One offers advantages, particularly for high-throughput applications. For high-throughput experiments, load peptides onto Evotips (instead of C18 StageTips; *see* Subheading 3.4 and **Note 8**). Insert the Evotips into the Evosep One system and connect the Evosep directly to the Orbitrap Astral for peptide separation and LC-MS/MS analysis. Its automated workflow reduces manual handling, increases reproducibility, and simplifies the sample loading process by using prepacked Evotips. Moreover, it allows efficient sample batching, making it especially beneficial in large-scale clinical or translational proteomic studies.

11. This CV was selected because it provided the highest number of identified protein groups during a titration curve of CVs tested on the same diluted sample (1:10 dilution of a 75,000 μm^2 sample, $N = 2$) (Fig. 4). In our hands, the best results were obtained with a CV of −53 combined with a carrier gas flow of 3.5. Users should be aware that the optimal carrier gas setting may depend on the chromatographic flow rate; for example, at 200 nL/min, values between 3.5 and 3.8 are generally suitable. Nevertheless, we recommend optimizing both the CV value and carrier gas setting for each instrument and chromatographic setup as these parameters can vary and must be determined on an instrument-by-instrument basis.
12. For low-input samples, we recommend Ionoptics Aurora Ultimate columns due to their smaller particle size (1.7 μm vs 2 μm), which delivers higher resolution and improved sensitivity compared to other commercially available columns. Although particle size is not directly related to sample amount, the improved separation efficiency of 1.7 μm particles is particularly advantageous for detecting low-abundance peptides in limited-input samples. In our laboratory practice, we consistently observe that smaller particle sizes yield sharper chromatographic peaks, which improves peptide separation and enhances the detection of low-abundance peptides. The drawback is that smaller particles generate higher back pressure, requiring the use of elevated operating temperatures to ensure stable flow and reproducible performance.
13. Free software packages for DIA data analysis, such as DIA-NN and FragPipe, can be used as alternatives to commercial platforms like Spectronaut, thus being particularly suitable for groups without access to commercial licenses. Both DIA-NN and FragPipe are actively developed, widely adopted in the proteomics community, and provide robust algorithms for directDIA analysis, spectral library generation, and statistical validation. These open-source solutions offer flexible and cost-effective options for DIA data analysis.
14. In our laboratory practice, we have found that these tools deliver results comparable to Spectronaut for many datasets although they may require more manual optimization of parameters and familiarity with open-source environments. Users are encouraged to evaluate both commercial and free solutions according to their project needs, computational resources, and desired level of automation.

Acknowledgments

The authors also acknowledge Styliani Koutsouli for helpful discussion on laser microdissection.

References

1. Mund A, Coscia F, Kriston A et al (2022) Deep visual proteomics defines single-cell identity and heterogeneity. Nat Biotechnol 40(8): 1231–1240
2. Makhmut A, Qin D, Fritzsche S et al (2023) A framework for ultra-low-input spatial tissue proteomics. Cell Syst 14(11):1002–1014 e1005
3. Muller T, Cremonini MA, Kliewer G et al (2023) Automated sample preparation for mass spectrometry-based clinical proteomics. Methods Mol Biol 2718:181–211
4. Muller T, Kalxdorf M, Longuespee R et al (2020) Automated sample preparation with SP3 for low-input clinical proteomics. Mol Syst Biol 16(1):e9111
5. Hughes CS, Moggridge S, Muller T et al (2019) Single-pot, solid-phase-enhanced sample preparation for proteomics experiments. Nat Protoc 14(1):68–85

Chapter 9

Laser Capture Microdissection Followed by Histone H1 Variant Analysis by Mass Spectrometry

Tiziana Bonaldi and Roberta Noberini

Abstract

The histone H1 family comprises of essential components of chromatin, which bind to the linker DNA connecting individual nucleosomes and contribute to the formation of higher-order chromatin structures. Multiple histone H1 exist, each with distinct interactions with the nucleosome that specifically influence chromatin organization and nuclear functions. Histone H1 variants have been shown to play a role as drivers in cancer and may serve as biomarkers for patient stratification. To overcome the limitations associated with antibody- and RNA-based methods for analyzing histone H1, we developed a mass spectrometry (MS)-based label-free approach to simultaneously analyze all somatic histone H1 variants in patient-derived samples. Here, we describe how this method can be used for the analysis of low-amount clinical samples obtained through laser capture microdissection of tissue sections.

Key words Histone H1 variants, Laser microdissection, Mass spectrometry, Epigenetics

1 Introduction

The major protein component of chromatin is represented by histones, which comprise core histones (H2A, H2B, H3, and H4) and the linker histone (histone H1). The core histones are assembled in a histone octamer, around which roughly 146 bp of DNA are wrapped, forming a nucleosome structure. Linker histone H1 binds the free DNA (~20 bp long) connecting individual nucleosomes and contributes to the establishment of higher-order chromatin structures [1]. Linker H1 comprises 11 variants in human and mouse: seven somatic variants (H1.0, H1.x, and H1.1 to H1.5), one oocyte-specific variant (H1oo), and three testis-specific variants (H1t, H1T2, and HILS1). Histones H1.1–1.5 are expressed in a replication-dependent manner, while histones H1.0 and H1x are transcribed during the course of the whole cell cycle [2]. Histone H1 variants can interact differently with the nucleosome, contributing to the formation of distinct chromatin

Roberta Noberini (ed.), *Laser Capture Microdissection: Methods and Protocols*, Methods in Molecular Biology, vol. 3015, https://doi.org/10.1007/978-1-0716-5154-4_9,

structures and regulating specific nuclear functions, which include DNA replication and repair, transcription, and genome stability (reviewed in [1]). Total histone H1 and H1 variant levels are often dysregulated in cancer (reviewed in [3]). Usually, the expression of replication-dependent variants is increased in tumor compared with normal tissues due to increased cell proliferation. In contrast, H1.0 is usually highly expressed in normal tissues and reduced in tumors, showing high intratumoral heterogeneity especially in aggressive and undifferentiated tumors [3].

H1 investigation in clinical samples relied traditionally on the analysis of RNA expression levels or antibody-based detection of protein levels (e.g., by immunochemistry). However, these methods are not ideal as alterations in mRNA levels may not correspond to changes in protein abundance and variant-specific antibodies may be challenging to obtain due to the high similarity of the H1.1–H1.5 variants. To overcome these issues, we developed a label-free mass spectrometry (MS)-based proteomics approach for the analysis of histone H1 variants in clinical samples, which has been applied also to laser-microdissected tissues containing down to 1000 cells [4].

Here, we will describe in detail this optimized protocol, which includes laser microdissection, protein extraction, in-gel digestion, and data analysis.

2 Material

2.1 Tissue Staining

1. Microscope slides and coverslips.
2. Microtome (for FFPE samples) or cryostat (for frozen samples).
3. Histolemon.
4. 95% and 70% ethanol (EtOH) solutions.
5. Hematoxylin solution, Mayer's.

2.2 Laser Microdissection

1. LCM membrane slides (e.g., polyethylene naphthalate membrane (PEN) slides).
2. Laser microdissector (here, a LMD7 LCM system (Leica Microsystems) was used).
3. Standard thin-walled 0.5 ml microcentrifuge tubes.

2.3 Protein Extraction

1. 95%, 50%, and 2% EtOH solutions.
2. Extraction buffer (2% SDS, 20 mM Tris–HCl, pH 7.4).
3. Gel loading 200 μL micropipette tips.
4. Tabletop centrifuge.

5. Bioruptor sonication device (Diagenode).
6. Heating block.

2.4 SDS-PAGE

1. 4–12% polyacrylamide gels.
2. Gel loading buffer (e.g., NuPAGE LDS Sample Buffer, Invitrogen), supplemented with 25 mM dithiothreitol.
3. BenchMark pre-stained protein ladder.
4. Gel running buffer (e.g., NuPAGE Mes SDS Running Buffer, Invitrogen).
5. Instant Blue Colloidal Blue Staining solution.
6. Gel running apparatus (e.g., XCell SureLock™ Mini-Cell, Invitrogen).

2.5 Trypsin In-Gel Digestion

1. LC-MS grade acetonitrile (ACN).
2. LC-MS grade EtOH.
3. Ammonium hydrogen carbonate (ABC).
4. Destaining buffer: 25 mM NH4HCO3 and 50% EtOH. Store at room temperature.
5. Digestion buffer: 50 mM ABC in water (pH 8.0). Store at room temperature.
6. Reduction buffer: 10 mM dithiothreitol (DTT) in 50 mM ABC. Store in small aliquots at −20 °C.
7. Alkylation buffer: 55 mM iodoacetamide in 50 mM ABC. Store in small aliquots at −20 °C and keep in the dark.
8. Trypsin solution: 12.5 ng/μL sequencing grade trypsin in 50 mM ABC. Prepare immediately prior to use and keep always on ice to minimize autocatalysis. Store at −20 °C.
9. Extraction solution: 3% trifluoroacetic acid and 30% ACN. Store at room temperature.
10. C18 StageTips: Empore SPE C18 disks.
11. SpeedVac vacuum centrifuge.

2.6 High-Performance Liquid Chromatography (HPLC)-Mass Spectrometry

1. HPLC solvent A: 0.1% formic acid in ddH2O.
2. HPLC solvent B: 0.1% formic acid/80% ACN in ddH2O.
3. EASY-nLC 1200 HPLC system (Thermo Fisher Scientific).
4. EASY-Spray column (Thermo Fisher Scientific), 25 cm long (inner diameter 75 μm, PepMap C18, 2 μm particles).
5. Q Exactive mass spectrometer (Thermo Fisher Scientific).
6. EASY-Spray™ Ion Source (Thermo Fisher Scientific).

2.7 Data Acquisition and Analysis

1. MaxQuant software.
2. PC with 2 GB RAM and a 64-bit Windows operating system.

3 Methods

First, the tissue area of interest are identified through the histopathological evaluation of a reference section stained with hematoxylin and eosin (Fig. 1a). These areas are then located on tissue sections mounted on membrane-coated LCM slides by comparing them with the reference section and are subsequently isolated using LCM. Whole protein extracts are then separated by SDS-PAGE, which allows removing MS contaminants, such as SDS and OCT traces that may be present in the protein extracts. This step also enables the enrichment of histone H1 by excising a gel band centered on its molecular weight. Proteins within the gel band are digested in-gel using trypsin and the resulting peptides are separated by liquid chromatography (LC) and analyzed by mass spectrometry (MS). The identification and quantification of histone H1 variants exploits the existence of peptides unique to each variant sequence and is based on the activation of the label-free quantification (LFQ) algorithm MaxLFQ [5].

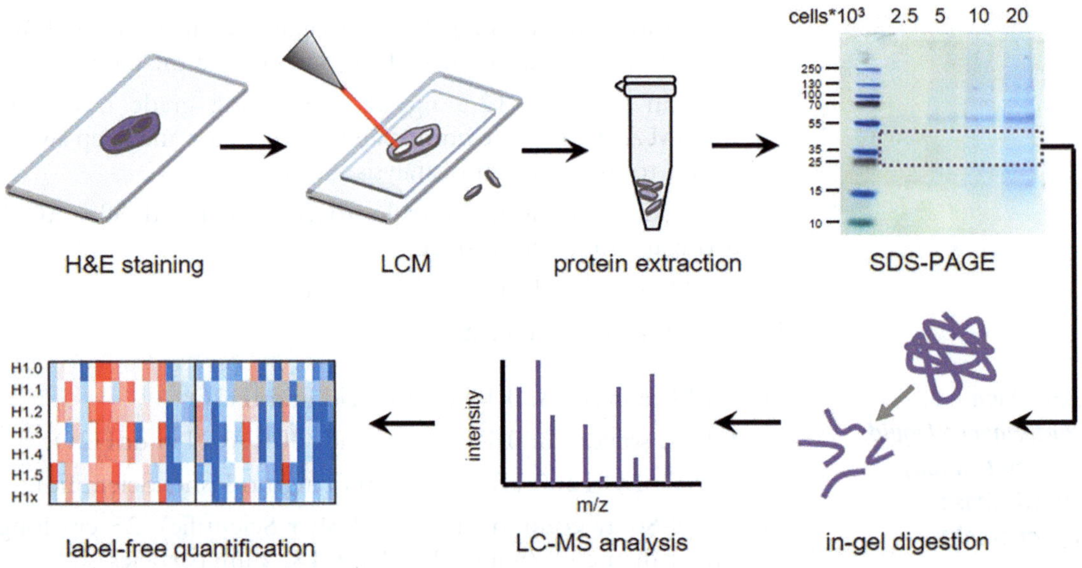

Fig. 1 Workflow of the protocol for the MS-based analysis of histone H1 variants from laser microdissected tissues. Regions of interest are identified and isolated by LCM. Proteins are extracted and loaded on a polyacrylamide gel, where a large band (20–45 kDa) around the size of histone H1 variants is excised for in-gel digestion. Digested peptides are analyzed by MS and histone H1 variants are quantified by label-free quantification

3.1 Tissue Staining and Laser Capture Microdissection (LCM)

1. Mount one 5 μm thick tissue section on a glass slide and stain the tissue with H&E with a standard staining protocol (*see* **Note 1**).
2. Identify the regions of interest by pathological evaluation (*see* **Note 2**).
3. Mount 10 μm thick tissue sections on LCM slides, which contain a membrane that aids tissue cutting (*see* **Notes 3** and **4**).
4. For FFPE samples only: To deparaffinize the sections, incubate them in histolemon twice for 5 min.
5. Rehydrate the sections in 100%, 95%, and 70% EtOH (two 3 min incubations for each EtOH concentration) and rinse the sections in distilled water for 30 s.
6. Stain the sections with the hematoxylin solution for 10 s and wash them with running tap water for 1 min.
7. Dry the sections under a chemical hood.
8. Use the pathologically annotated section from point 1 as reference to identify the areas of interest (*see* **Note 5**).
9. Place the LCM slides with the tissue face down on the slide holder in a laser microdissector.
10. Cut the areas of interest. Typical settings using the UV-based LMD7 LMD instrument used in [4] are (*see* **Notes 6** and 7):
    - "Draw and cut" mode.
    - Wavelength: 349 nm.
    - Power: 42.
    - Aperture: 6.
    - Speed: 5.
    - Offset: 80%.
11. Collect the areas of interest in 0.5 mL plastic tubes and store them at 4 °C.

3.2 Protein Extraction

1. Centrifuge the tissue pieces in a tabletop centrifuge for 3 min at maximum speed to collect the tissue pieces at the bottom of the tube.
2. Open the tubes carefully to avoid losing tissue pieces. Add 200 μL of a 95% EtOH solution and incubate for 3 min at RT (*see* **Note 8**).
3. Centrifuge at 16,000 × g for 3 min at 4 °C and carefully discard the supernatant using a gel loader tip.
4. Rehydrate the tissue pieces by repeating **steps 2** and **3** with 50% and 20% EtOH solutions.

5. Add 40 μL of extraction buffer and homogenize the tissue pieces by sonication in a Bioruptor sonication device, with the following settings: 30 s on, 30 s off, 10 cycles, power: high, at room temperature (*see* **Note 9**).
6. For FFPE tissues only: de-crosslink the proteins with a 45 min incubation at 95 °C, followed by a 4-hour incubation at 65 °C.
7. Centrifuge the sample at 16,000 × g for 1 min at 4 °C, discard any remaining tissue pieces, and store the supernatant at −20 °C.

3.3 SDS-PAGE Separation and Trypsin In-Gel Digestion

1. Load all the tissue extracts on a 4–12% polyacrylamide gel. Prior to loading, add a gel loading buffer to the samples. Also, load a pre-stained protein ladder on the same gel. Run the gel under appropriate conditions until the proteins are properly separated.
2. Stain the gel with colloidal Coomassie staining, following the manufacturer's indications (*see* **Note 10**).
3. Excise a large band (20–45 kDa) centered around the molecular weight of histone H1, chop it into pieces of approximately 1 mm^3, and transfer the gel cubes to a 1.5 ml plastic tube (*see* **Notes 11** and **12**).
4. Destain the gel pieces in the destaining buffer. An excess solution should be used (250–500 μl, depending on band size). Discard the liquid and repeat until the staining is completely removed.
5. Dehydrate the gel pieces in 100% EtOH for 10 min at room temperature in agitation (1400 rpm) using a volume corresponding to approximately 3–4 times the total volume of the gel pieces. Repeat until the gel pieces are hard and white. Discard the liquid afterward and dry the sample in a vacuum centrifuge for 5 min.
6. Rehydrate the gel pieces in the reduction buffer and incubate them for 60 min at 56 °C, in agitation at 300–400 rpm. Ensure that the gel pieces are fully covered with the solution (*see* **Note 13**). After the incubation, discard the liquid.
7. Add the alkylation buffer and incubate the gel pieces in the dark for 45 min at room temperature. Ensure that the gel pieces are fully covered with the solution (*see* **Note 13**). After the incubation, discard the liquid.
8. Wash the gel pieces with the digestion buffer for 20 min at 25 °C (1,400 rpm), and dehydrate them as described in step 5. Repeat twice.
9. Rehydrate the gel pieces in the trypsin solution at 4 °C for 20–30 min using enough solution to cover the dehydrated gel pieces (*see* **Note 13**). Once the gel pieces are completely

rehydrated, remove the excess trypsin solution and cover the gel pieces with the digestion buffer. Incubate the sample overnight at 37 °C.

10. The day after, transfer the supernatant to a fresh 1.5 mL tube.
11. Extract the peptides from the gel by incubating the gel pieces twice with the extraction buffer, and twice with 100% ACN for 10 min at room temperature, in agitation at 1400 rpm. After each extraction cycle, recover the supernatant and combine it with that from step 10.
12. Concentrate the samples in a vacuum centrifuge to reach 5–10% of the original volume.
13. Dilute the samples in Solvent A and load them on C18 StageTips [6] (*see* **Note 14**). Digested peptides can be stored on the StageTips at 4 °C until use.

3.4 LC-MS/MS

Liquid chromatography enables the separation of peptides generated from histone digestion prior to their analysis using a high-resolution mass spectrometer. Here, we provide a brief description of the parameters used with an EASY-nLC 1200 HPLC system, connected via an EASY-Spray column to a Q Exactive Plus Orbitrap mass spectrometer. While this setup is used as an example, other liquid chromatography and mass spectrometry configurations can also be employed, provided they offer comparable sensitivity and resolution.

1. Elute the peptides from the StageTips using 40 μL of solvent B, by placing the StageTip in a fresh 1.5 plastic tube and centrifuging at 1200 × g for approximately 5 minutes, or until all the solution passed through the C18 resin.
2. Dry the peptides in a vacuum centrifuge and resuspend them in 5 μL of solvent A.
3. Load 3 μL of the eluted peptides onto a C_{18} HPLC column at constant flow of 500 nl/min in solvent A. After sample loading, apply a 95 min 3%–60% gradient of solvent B (80 min 3–30%, 10 min 30–40%, 5 min 40–60%), at a flow rate of 250 nL/min, for peptide separation.
4. For mass spectrometry acquisition on a Q Exactive HF instrument, operate the mass spectrometer in the data-dependent acquisition mode (DDA) to automatically switch between full scan MS and MS/MS acquisition using the Xcalibur software package.

   Typical acquisition settings:

   - Full scans: scan range: *m/z* 300 to 1350; resolution: 70000, AGC target: 3×10^6, maximum injection time: 20 ms.

- MS/MS: resolution: 17500; AGC target: 1×10^5; maximum injection time: 80 ms; loop count:12.
- General settings: electrospray voltage: 1.7 kV; no sheath and auxiliary gas flow; ion transfer capillary temperature: 250 ° C; dynamic exclusion: 20 s; normalized collision energy: 28%.

3.5 Data Analysis

All the somatic histone H1 variants can be distinguished and quantitated, thanks to the presence of at least one unique peptide for each sequence, both in mouse and in human. Figure 2 shows the list of unique peptides identified by MS in mouse pancreas whole sections or laser-microdissected areas containing 1000–20,000 cells.

1. Import the acquired raw data in the MaxQuant software [4] and write the experimental design.
2. Load the FASTA sequence file from the appropriate organism in the "Sequence" tab. Databases can be downloaded from the UniProt (https://www.uniprot.org/) or Ensemble websites (https://useast.ensembl.org/index.html) (*see* **Note 15**).
3. Most settings are left at default, including enzyme specificity (trypsin), maximum missed cleavages (2), variable modifications (methionine oxidation and N-terminal acetylation), and false discovery rate (FDR, 1%).
4. Activate the label-free software MaxLFQ [5] (LFQ minimum ratio count: 1) and the "match between runs" feature (match time window: 2 min) and launch the search.
5. Once the search is finished, extract the LFQ values for histone H1 variants from the "protein groups" MaxQuant output file and use them for comparative analyses after log transformation (*see* **Note 16**).

4 Notes

1. The protocol can be applied to FFPE, OCT-frozen, and fresh-frozen samples. To cut sections from fresh-frozen samples, the tissue can be immobilized on the cryostat through a drop of frozen water.
2. Molecular profiling approaches, such as multi-antibodies staining imaging methods or MALDI imaging [7], can be used as an alternative to morphological evaluation to identify the tissue areas of interest.
3. PEN (polyethylene naphthalate) LCM slides should be activated for 30 minutes under UV light prior to use. PPS (polyphenylene sulfide) membranes do not need any activation step.

Histone	Peptide sequence	whole slice		20000 cells		5000 cells		2500 cells		1000 cells	
		FFPE	OCT	FFPE	OCT	FFPE	OCT	FFPE	OCT	FFPE	OCT
H1.0	GVGASGSFR										
	LVTTGVLK										
	LVTTGVLKQTK										
	QTKGVGASGSFR										
	RLVTTGVLK										
	TENSTSAPAAK										
	TENSTSAPAAKPK										
	VGENADSQIK										
	YSDMIVAAIQAEK										
	H1.0 variant quantified										
H1.1	KKPAGPSVSELIVQAVSSSK										
	KPAGPSVSELIVQAVSSSK										
	KSLAAAGYDVEK										
	PAGPSVSELIVQAVSSSK										
	SETAPVAQAASTATEK										
	SETAPVAQAASTATEKPAAAK										
	SLAAAGYDVEK										
	H1.1 variant quantified										
H1.2	GILVQTK										
	KPAAAAVTK										
	SEAAPAAPAAAPPAEK										
	SLVSKGILVQTK										
	H1.2 variant quantified										
H1.3	SETAPAAPAAPAPVEK										
	H1.3 variant quantified										
H1.4	KTSGPPVSELITK										
	SETAPAAPAAPAPAEK										
	TSGPPVSELITK										
	TSGPPVSELITKAVAASK										
	H1.4 variant quantified										
H1.5	ALAAGGYDVEK										
	ATGPPVSELITK										
	GGVSLPALK										
	KALAAGGYDVEK										
	KATGPPVSELITK										
	SETAPAETAAPAPVEK										
	H1.5 variant quantified										
H1x	ALVQNDTLLQVK										
	YSQLVVETIR										
	H1.1 variant quantified										

Fig. 2 Histone H1 variant analysis from laser-microdissected samples. Unique peptides with up to two miscleavages identified from OCT-frozen or FFPE mouse pancreas whole sections, or laser-microdissected areas corresponding to 1000–20,000 cells are marked in blue. Peptides identified in less than three samples are not displayed. Orange indicates whether histone H1 variants were quantified through the MaxLFQ algorithm, based on one or more unique peptides

4. The number of sections to be cut must be defined based on the size of the area of interest and the amount of histone H1 variant present in the sample, as well as the MS instrumentation. With the described MS setup, 1000 cells were tested as the minimum input amount. However, with more recent instrumentation, this cell number can be likely substantially reduced.

5. For tissue areas difficult to visualize, the sections mounted on LCM slides may need to undergo pathological evaluation to correctly identify the regions of interest.
6. Areas that are sufficiently large and homogeneous can also be isolated by manual macrodissection, by scraping the tissue off a regular glass slide with a scalpel into a plastic tube.
7. The cutting conditions must be optimized based on the specific specimen in use. The described settings were used for mouse pancreatic tissues and breast cancer human samples.
8. At this point, the solution with the tissue pieces may be transferred to a 1.5 mL plastic tube, which is more convenient to use for the following steps.
9. If large tissue pieces remain after the first round of sonication, the procedure should be repeated.
10. For particularly low sample amounts, protein extracts may not be visible on the Coomassie-stained gel.
11. The goal of digesting only proteins within a specific molecular weight range is to enrich for histone H1 while still retaining enough background proteins to enable reliable normalization of protein levels within the sample.
12. We also developed another MS-based method enabling the quantitation of core histone post translational modifications (PTMs) from laser-microdissected tissue areas containing 1000 cells [7]. This method can be performed in parallel with the one described here to analyze histone H1 variants, as well as core histone PTMs from the same sample loaded onto the gel, as the two approaches involve digestion of gel bands corresponding to different molecular weight ranges.
13. The volume depends on the size of the gel cubes and must be determined case by case.
14. StageTips are reversed-phase chromatography micro-columns, which are manually assembled by placing two 14-gauge StageTip plugs of C18 resin in a 200 μl pipette tip. Prior to being used, StageTips must be activated with 100% methanol, washed with solvent B, and equilibrated twice with solvent A.
15. New sequence databases should be configured in the "Configuration" tab.
16. More detailed information on the identified peptides can be obtained from the "*evidence*" or "*peptides*" output files, which report all identified LC-MS features and peptide sequences, respectively.

Acknowledgments

T. B. research activity is supported by grants from the Italian Association for Cancer Research (grant# IG-2023-28767).

References

1. Fyodorov DV, Zhou BR, Skoultchi AI et al (2018) Emerging roles of linker histones in regulating chromatin structure and function. Nat Rev Mol Cell Biol 19(3):192–206
2. Zlatanova J, Doenecke D (1994) Histone H1 zero: a major player in cell differentiation? FASEB J 8(15):1260–1268
3. Scaffidi P (2016) Histone H1 alterations in cancer. Biochim Biophys Acta 1859(3):533–539
4. Noberini R, Morales Torres C, Savoia EO et al (2020) Label-free mass spectrometry-based quantification of linker histone H1 variants in clinical samples. Int J Mol Sci 21(19)
5. Cox J, Hein MY, Luber CA et al (2014) Accurate proteome-wide label-free quantification by delayed normalization and maximal peptide ratio extraction, termed MaxLFQ. Mol Cell Proteomics 13(9):2513–2526
6. Rappsilber J, Mann M, Ishihama Y (2007) Protocol for micro-purification, enrichment, pre-fractionation and storage of peptides for proteomics using StageTips. Nat Protoc 2(8): 1896–1906
7. Noberini R, Savoia EO, Brandini S et al (2021) Spatial epi-proteomics enabled by histone post-translational modification analysis from low-abundance clinical samples. Clin Epigenetics 13(1):145

Chapter 10

A Spatial Proteomics Approach to Isolate and Analyze Amyloid Proteins Using Laser Capture Microdissection and Mass Spectrometry

Jennifer Aguilan, Maxwell Horton, Jeffrey E. Pessin, and Simone Sidoli

Abstract

Amyloidosis comprises a heterogeneous group of disorders marked by the extracellular deposition of insoluble protein fibrils, known as amyloids, which can disrupt normal tissue architecture and lead to organ dysfunction. Accurate identification and subtyping of the amyloidogenic protein are critical for clinical management as treatment strategies vary significantly depending on the underlying protein species. While conventional diagnostic tools such as Congo red staining and immunohistochemistry are commonly used, they suffer from limited specificity and antibody availability, often resulting in misclassification. Mass spectrometry (MS)-based proteomics has become the gold standard for amyloid subtyping, offering unmatched sensitivity and proteome-wide coverage.

Here, we describe a spatial proteomics protocol that integrates laser capture microdissection (LMD) with advanced LC-MS/MS acquisition methods, including data-dependent acquisition (DDA), data-independent acquisition (DIA), and High-Field Asymmetric Waveform Ion Mobility Spectrometry (FAIMS). With the integration of these methods, we present the analysis of formalin-fixed paraffin-embedded (FFPE) tissue sections. This comprehensive workflow enables precise excision of amyloid-rich regions and enhances detection of amyloid proteins and co-deposited biomarkers from minimal tissue input. The combination of FAIMS with DIA and DDA not only improves the depth of proteomic coverage but also increases reproducibility and sensitivity, making it particularly suitable for low-abundance samples. This protocol provides a robust and scalable platform for the accurate molecular subtyping of amyloidosis and has the potential to inform personalized therapeutic decisions in clinical pathology.

Key words Amyloidosis, MALDI-MSI, Laser capture microdissection, FAIMS, DIA, DDA, Amyloid protein subtyping

1 Introduction

Amyloidosis is a heterogeneous group of protein misfolding diseases characterized by the extracellular deposition of insoluble amyloid fibrils, which disrupt tissue architecture and lead to

Jennifer Aguilan and Maxwell Horton contributed equally with all other contributors.

Roberta Noberini (ed.), *Laser Capture Microdissection: Methods and Protocols*, Methods in Molecular Biology, vol. 3015, https://doi.org/10.1007/978-1-0716-5154-4_10,

progressive organ dysfunction [1]. There are over 36 known amyloidogenic proteins in humans, each designated with an "A" prefix (e.g., AL for immunoglobulin light chain, ATTR for transthyretin). The clinical manifestation depends on the subtype and affected organs. The most involved are the heart, kidneys, liver, and nervous system [2]. Accurate and early subtyping of amyloid proteins is essential for effective management as each subtype requires a distinct therapeutic approach [3]. For example, AL amyloidosis is treated with chemotherapy targeting plasma cells, ATTR amyloidosis may be managed with TTR (transthyretin) stabilizers or liver transplantation, and AA (serum amyloid A) amyloidosis requires control of chronic inflammation [4].

Initial diagnosis often begins with Congo red staining, which identifies amyloid deposits based on apple-green birefringence under polarized light. While this technique is sensitive to detecting fibrils, it lacks specificity for the amyloid subtype [2]. Immunohistochemistry (IHC) has traditionally been employed to identify amyloid types, but it suffers from key limitations: low sensitivity and specificity, epitope masking in formalin-fixed samples, incomplete antibody panels, and reliance on multiple serial sections [3]. Studies have shown that IHC-based subtyping can misclassify amyloid, leading to inappropriate therapies, especially when antibodies cross-react or fail to detect mutated proteins. Mass spectrometry (MS)-based proteomics has revolutionized amyloid diagnostics by enabling unbiased and highly sensitive identification of the fibril protein. A landmark study from the Mayo Clinic demonstrated near 100% accuracy using laser capture microdissection (LMD) of Congo red-positive areas followed by tandem mass spectrometry [3]. This approach is now considered the gold standard, and it has been validated across large patient cohorts and is widely adopted in reference laboratories [4].

The workflow involves LMD isolation of amyloid-rich tissue, protein extraction and tryptic digestion, and LC-MS/MS analysis. A schematic overview of the full spatial proteomics pipeline, from laser capture microdissection (LMD) to LC-MS/MS analysis and computational processing, is illustrated in Fig. 1. Using database searches, dominant fibril proteins are identified along with co-deposited molecules such as serum amyloid P component and apolipoproteins. This method is highly sensitive, can distinguish rare or mixed amyloid types, and detects amino acid variants, critical for hereditary forms like variant ATTR [4]. Unlike IHC, which tests only for selected proteins, MS surveys the entire proteome in a single assay, offering both breadth and precision.

Traditionally, data-dependent acquisition (DDA) [5] has been employed in amyloid proteomics. DDA selectively fragments the most intense precursor ions for sequencing, enabling high-confidence peptide identifications. However, it may miss low-abundance peptides, especially in complex or small samples,

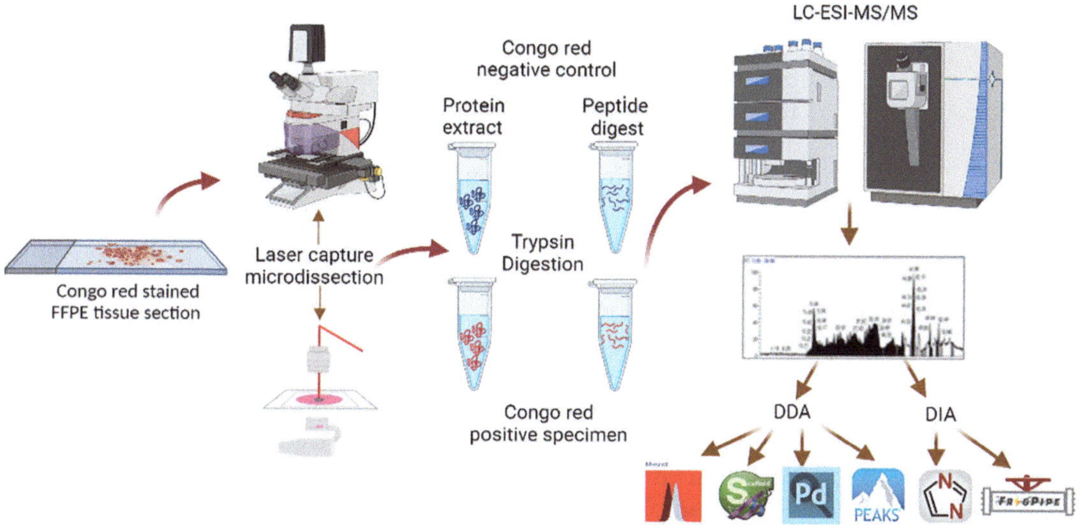

Fig. 1 Spatial proteomics workflow for amyloid protein identification using laser capture microdissection (LMD) and LC-MS/MS. Congo red-stained FFPE tissue sections are mounted and dissected using a Leica LMD system. Proteins are extracted, digested, and analyzed via LC-MS/MS with FAIMS, DIA, and DDA methods. Spectral data are interpreted using proteomics software including, for example, Mascot, Scaffold, Proteome Discoverer, PEAKS, DIA-NN, and FragPipe

due to its stochastic nature. To overcome this, data-independent acquisition (DIA) [6] has gained traction. DIA systematically fragments all ions within defined m/z windows, improving reproducibility and capturing low-abundance species. This results in more consistent peptide detection across samples. While DIA requires more advanced data processing (e.g., spectral libraries or deconvolution algorithms), its application to amyloid proteomics enhances coverage and quantification, especially in challenging FFPE tissues [7].

Recent improvements in ion separation technologies have further increased MS sensitivity. High-Field Asymmetric Waveform Ion Mobility Spectrometry (FAIMS) separates ions in the gas phase before MS analysis [8], filtering out chemical noise and enriching for multiply charged peptides. This has become very advantageous for complex and low-input samples, for example, single-cell proteomics [9]. In studies using FFPE tissue, FAIMS has improved signal-to-noise ratios and expanded protein identifications [10]. For amyloidosis, where laser-microdissected tissue yields limited material, FAIMS can reveal amyloid peptides that would otherwise remain undetected, enhancing diagnostic confidence.

MS-based proteomics has proved to provide substantial clinical impact. Specifically, for AL amyloidosis, proteomics routinely identifies dominant light-chain peptides (kappa or lambda), confirming the diagnosis and guiding chemotherapy [11]. MS can distinguish

AL from AH or mixed types, which IHC may misclassify. In ATTR amyloidosis, MS not only detects transthyretin but also identifies pathogenic variants (e.g., V30M, T60A), distinguishing hereditary from wild-type forms and supporting genetic counseling [4]. In AA amyloidosis, proteomics confirms serum amyloid A as the fibril protein, directing treatment toward inflammation control [4]. It avoids unnecessary therapies when monoclonal proteins are present but unrelated. Furthermore, proteomics has also uncovered under-recognized types like ALECT2 amyloidosis, which were frequently misdiagnosed before MS became available [12].

In summary, the integration of spatial proteomics techniques with advanced acquisition methods like DIA and FAIMS provides a robust, sensitive, and comprehensive workflow for amyloid protein subtyping. This approach outperforms traditional methods in both precision and scope, enabling confident diagnosis from limited FFPE samples. As these technologies become more accessible, mass spectrometry-based amyloid subtyping is poised to become routine in clinical pathology, supporting personalized therapeutic decisions and improved patient outcomes [3]. In this chapter, we provide a comprehensive workflow for amyloid protein identification, incorporating LMD/LC-MS/MS, FAIMS, DIA, and DDA optimized in a recent publication [13]. For complementary spatial analysis, matrix-assisted laser desorption ionization mass spectrometry imaging (MALDI-MSI) can be performed as shown in Fig. 2. These methods, when applied together, offer a robust approach for amyloidosis subtyping with higher sensitivity and accuracy than traditional techniques. By enabling precise identification of amyloid protein subtypes, this protocol supports better clinical decision-making, helping guide appropriate therapeutic strategies for amyloidosis patients.

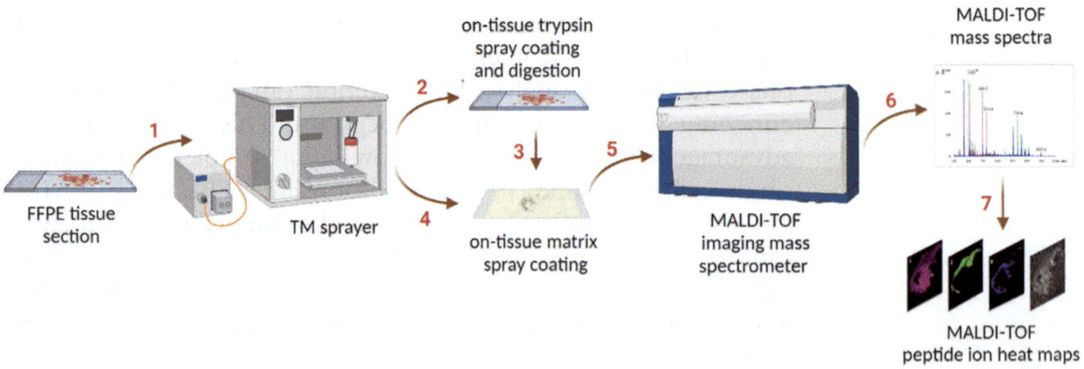

Fig. 2 MALDI mass spectrometry imaging (MSI) workflow from FFPE tissue. (1) Slides are heated to melt paraffin. (2) A sprayer deposits trypsin for in situ digestion. (3) Digestion proceeds on slide. (4) Matrix is sprayed. (5) MALDI-MSI is performed. (6) Ion maps are generated for m/z values. (7) Spatial distributions of proteins are visualized

2 Materials

2.1 Reagents

1. Methanol.
2. Xylene.
3. 100% ethanol.
4. Distilled water.

2.2 Buffers

1. 50 mM ammonium bicarbonate (NH_4HCO_3), pH 8.0.
2. 10 mM Tris–HCl, pH 8.0.
3. 60% acetonitrile (ACN), 0.1% trifluoroacetic acid (TFA).
4. Mix desalting resin: 50 mg/mL Oasis HLB resin in 100% HPLC-grade CAN.
5. nanoLC buffer A: 0.1% formic acid in HPLC-grade water.
6. nanoLC buffer B: 0.1% formic acid, 80% HPLC-grade ACN, in HPLC-grade water.

2.3 Solutions

1. Ethanol solutions, 95% and 70%. Dilute 100% ethanol in distilled water.
2. Dithiothreitol (DTT) solution, 5 mM. Dissolve DTT in 50 mM ammonium bicarbonate. Prepare fresh before use. Store at −20 °C in aliquots if needed.
3. Iodoacetamide (IAA), 20 mM. Dissolve iodoacetamide in 50 mM ammonium bicarbonate. Protect from light. Prepare fresh before use.
4. Trypsin solution, 20 ng/μL. Reconstitute sequencing-grade trypsin in 50 mM ammonium bicarbonate to a final concentration of 20 ng/μL. Store aliquots at −20 °C. Avoid repeated freeze-thaw cycles.
5. Congo red staining solution. Dissolve 0.5% Congo red in saturated aqueous sodium chloride solution. Add 1% sodium hydroxide. Filter before use. Prepare fresh or store at 4 °C protected from light for up to 1 week.
6. Hematoxylin solution. Use commercially available hematoxylin or prepare according to standard protocol. Filter before use.
7. Tissue extraction buffer: (i) 10 mM Tris–HCl, pH 8.0, (ii) 2 mM EDTA, (iii) 0.002% Zwittergent 3–16. Mix all components, adjust pH, and filter-sterilize. Store at 4 °C.
8. Desalting Resin Suspension: 50 mg/mL Oasis HLB resin suspended in 100% HPLC-grade acetonitrile. Vortex or stir thoroughly before dispensing.

2.4 Equipment

1. Microtome.
2. Laser Capture Microscope LMD6 (Leica Microsystems).
3. TM Sprayer for Trypsin Coating (HTX Technologies).
4. Orbitrap Exploris 480 with FAIMS coupled online with a Dionex Ultimate 3000 HPLC (Thermo Fisher Scientific).
5. 1.5 mL microcentrifuge tubes.
6. 0.2 mL PCR tube.
7. Polyethylene naphthalate (PEN) membrane slides 4 μm.
8. pH indicator strips (pH 0–14).
9. 96-well polypropylene filter plate containing 10 μm PE frit.
10. 96-well full-skirt PCR microplate.
11. Pipettes from P10 to P1,000 range with respective tips.
12. –80° C freezer.
13. Water bath sonicator.
14. Refrigerated centrifuge.
15. Vortex.
16. SpeedVac.
17. 96-well vacuum manifold.
18. Silica capillary with 75 μm internal diameter (ID).
19. HPLC glass vial.
20. C_{18}-AQ reversed-phase 3 μm particles.
21. Micro-stir magnet.
22. Pressure bomb.
23. 300 μm ID s 0.5 cm C18 trap column.
24. Reprosil-Pur C18-AQ (3 μm) analytical nano-column packed in-house.

3 Methods

3.1 FFPE Tissue Sample Preparation

Formalin-fixed paraffin-embedded (FFPE) tissue samples are commonly used in clinical and research settings for their excellent preservation of cellular and tissue architecture. However, the paraffin embedding and formaldehyde cross-linking present challenges for downstream proteomic analysis. Deparaffinization and rehydration of the tissue, followed by specific staining, are essential steps to prepare the samples for laser capture microdissection (LMD) and mass spectrometry analysis. This section details the procedures for preparing FFPE tissue sections, including the removal of paraffin and staining for amyloid deposits, using Congo red.

3.1.1 Deparaffinization

1. Prepare 10 μm sections of formalin-fixed paraffin-embedded (FFPE) tissue using a microtome.
2. Mount the sections on PEN membrane slides designed for LMD.
3. Incubate the slides in an oven at 60 °C for 1.5 hours to melt the paraffin.
4. Submerge the slides in xylene for 10 minutes, twice, to completely dissolve the paraffin.
5. Wash the slides in a series of ethanol solutions to rehydrate the tissue sections:
6. Immerse in 100% ethanol for 5 minutes, twice.
7. Immerse in 95% ethanol for 3 minutes.
8. Immerse in 70% ethanol for 3 minutes.
9. Rinse the slides in distilled water to remove any residual ethanol.

3.1.2 Congo Red Staining and Hematoxylin Counterstaining

1. Stain the tissue sections in freshly prepared Congo red solution.
2. Prepare 0.5% Congo red in saturated sodium chloride solution.
3. Add 1% sodium hydroxide to enhance the staining process.
4. Immerse the tissue sections in Congo red solution for 30 minutes at room temperature, ensuring even staining.
5. Rinse the slides in distilled water to remove excess stain.
6. Counterstain the sections with hematoxylin for 3–5 minutes to visualize cell nuclei.
7. Rinse the slides again in distilled water to clear excess hematoxylin.
8. Dehydrate the stained sections by dipping them briefly into increasing concentrations of ethanol (70%, 95%, and 100%) for 1 minute each.
9. Let the slides air-dry for LMD.

3.2 Laser Capture Microdissection (LMD)

LMD is a precise method used to isolate specific regions of interest from tissue sections, such as amyloid-positive areas stained with Congo red. The accuracy of this technique allows for targeted extraction of protein-rich areas, ensuring that only the relevant tissue is collected for subsequent analysis. In this section, we describe the process of identifying amyloid deposits under a microscope and dissecting them using a laser-based system, preparing the samples for protein extraction and downstream mass spectrometry.

3.2.1 Selection of Congo Red-Positive Areas

1. Place the dried, stained PEN membrane slides under a Leica LMD6 Laser Capture Microscope.

2. Identify Congo red-positive areas of amyloid deposits using bright-field microscopy (*see* **Note 1**).
3. Use Texas red filter to confirm the presence of amyloid fibrils, which will exhibit bright red color using fluorescence microscopy.
4. Use hematoxylin-stained nuclei as a guide to focus the microscope and accurately identify relevant tissue areas.

3.2.2 Laser Microdissection

1. Adjust the laser settings on the Leica ASLMD system to focus and cut around the Congo red-positive regions.
2. Carefully perform the dissection by activating the laser and collecting the tissue sections into the cap of a 0.2 mL PCR tube preloaded with 20 μL of tissue extraction buffer.
3. Ensure that enough tissue is captured to provide sufficient protein for analysis (*see* **Note 2**).
4. Check the collection under the microscope to ensure the accuracy of dissection and collection.

3.3 Protein Extraction

The extraction of proteins from FFPE tissue sections is a critical step in preparing samples for proteomic analysis. The process begins with breaking the formaldehyde-induced cross-links to release proteins, followed by sonication to enhance extraction. Once proteins are liberated, enzymatic digestion with trypsin cleaves them into peptides suitable for mass spectrometry. This section outlines the detailed steps for protein extraction, digestion, and peptide purification, ensuring that the samples are ready for LC-MS/MS analysis.

1. After LMD, place the PCR tubes containing dissected tissue in a heating block set to 95–98 °C for 90 minutes to break cross-links formed during formalin fixation.
2. Following heating, sonicate the tubes for 1 hour at 80 °C to enhance tissue disruption and protein extraction.
3. Centrifuge the tubes at maximum speed (14,000 g) for 10 minutes to remove any insoluble debris.
4. Carefully transfer the supernatant containing extracted proteins to a new 0.2 mL PCR tube.
5. Bring the volume up to 25 μL.

3.4 Protein Reduction, Alkylation, and Trypsinization

To achieve optimal peptide generation for mass spectrometry analysis, proteins must first undergo reduction and alkylation to break disulfide bonds and prevent reformation, ensuring complete unfolding of the protein structures. This process enhances trypsin accessibility, allowing efficient proteolytic cleavage. Trypsin, a protease with high specificity for lysine and arginine residues, digests the proteins into peptides suitable for analysis. In this section, we detail the steps required to reduce and alkylate the proteins, followed by an overnight trypsinization to generate peptides for downstream LC-MS/MS analysis.

1. Add 2 μL of 50 mM DTT in 10 mM Tris–HCl to a final concentration of 5 mM DTT.
2. Incubate for 65 °C for 30 min.
3. Add 1 uL of 400 mM iodoacetamide in 10 mM Tris–HCl to a final concentration of 15 mM iodoacetamide.
4. Incubate for 15 min at room temperature, protected from light.
5. Prepare a solution of 100 ng/μl trypsin in 10 mM Tris–HCl.
6. Add 5 μL (500) ng of trypsin solution to each sample.
7. Digest overnight at room temperature.
8. Add 0.5 uL of 50% TFA, stop the digestion, and acidify the solution.
9. Dry the pooled volume in the SpeedVac.

3.5 Sample Desalting

Desalting allows for the removal of residual salts and contaminants in a sample, which can adversely affect protein function and stability and can cause column clogging and/or instrument contamination during HPLC-MS analysis. Samples are passed through a column packed with a reversed-phase chromatography resin to allow for efficient separation of proteins of interest.

1. Mix desalting resin on a magnetic stir plate without heat.
2. A 96-well collection plate should be placed to collect flow-through.
3. Add 70 μL of resin suspension to each well of a 96-well polypropylene filter plate containing 10 μm PE frit.
4. Turn on vacuum gently; discard the flow-through.
5. Wash resin with 100 μL 0.1% TFA.
6. Turn on vacuum gently; discard the flow-through.
7. Resuspend samples in 100 μL 0.1% TFA (pH should be 2–3).
8. Load samples onto filter plate.
9. Turn on vacuum gently; discard the flow-through.
10. Wash samples with 100 μL 0.1% TFA.
11. Turn on vacuum gently; discard the flow-through.
12. Replace the 96-well collection plate with a new 96-well full-skirt PCR microplate.
13. Add 60 μL 60% acetonitrile/0.1% TFA to filter plate wells.
14. Turn on vacuum gently to elute.

15. Collect flow-through and dry in speed vacuum (samples can be stored at −80 °C at this stage).
16. Resuspend samples in 10 μL 0.1% TFA.
17. Proceed with a proteomics data acquisition.

3.6 C_{18} Nano-Liquid Chromatography Mass Spectrometry (nLC-MS) Analysis

Here, we present a widely adopted workflow to analyze proteins using nano-liquid chromatography coupled online with tandem mass spectrometry (nLC-MS/MS). Notably, other approaches for identifying the protein eluate are also adaptable, including other MS-based configurations and also methods not utilizing MS, for example, SomaScan, Olink, etc. For non-MS methods, please skip Subheadings 3.5 and 3.6. To generate the data for the reference manuscript [13], we utilized a nLC-MS/MS system using a two-column system setup with a 300 μm ID × 0.5 cm C18 trap column and a 75 μm ID × 25 cm Reprosil-Pur C18-AQ (3 μm) analytical nano-column packed in-house. **Steps 1–8** can be omitted if using commercial columns.

1. Cut ~30 cm of fused silica capillary with 75 μm internal diameter (ID).
2. Tape one end of the capillary on the bench.
3. Apply heat near the end of the capillary using a torch while gently pulling from the other end. A few seconds later, the capillary will elongate and detach from the bench creating a tip at the end.
4. Remove the "hair" part of the tip, for example, with the finger, until it looks like a tip for nano-columns.
5. Prepare the resin slurry in a clean HPLC glass vial for column packing, including C_{18}-AQ reversed-phase 3 μm particles, 100% methanol, and a micro-stir magnet. Place the resin slurry in a pressure bomb and turn on magnetic stirring.
6. Put the capillary in the pressure bomb, adjust the end height of capillary in the glass vial to make sure it does not touch the magnet stir, secure it, and open the gas tank (helium, nitrogen, or air) to apply high pressure pushing resin into the capillary.
7. Leave the pressure on until the capillary is packed with resin for at least ~25 cm.
8. Remove the column from the pressure bomb and connect it to the nLC.
9. Program the HPLC method as follows: from 4 to 34% buffer B in 120 min, from 34% to 90% B in 5 min, and 5 min at isocratic 90% B. Buffers A and B composition are described in Subheading 2.2 (#4 and #5, respectively).
10. Set flow rate to 300 nL/min.

11. Set up the MS acquisition method to perform data-dependent acquisition (DDA) and data-independent acquisition (DIA) to generate MS/MS spectra of the detected peptide signals.
12. Resuspend samples in 10 μL of 0.1% TFA prior placing the samples in the nLC autosampler.
13. Run the nLC-MS method as programmed for canonical proteomics acquisition (*see* **Note 3**).

3.7 Data Analysis

Raw files obtained from the nLC-MS/MS runs can be analyzed with different proteomics software available on the market or free of charge.

1. Import data into a peak detection software compatible with the MS platform used (*see* **Note 4**).
2. Select the proper database (human, mouse, etc.).
3. Set N-terminal acetylation as variable modification and carbamidomethyl cysteine as fixed modification.
4. Specify trypsin as the digestive enzyme with two missed cleavage sites allowed.
5. Set mass tolerance accordingly to the mass analyzer utilized for the spectra acquisition.
6. Export the analysis as a spreadsheet. In the next section, we will illustrate how to analyze data using Microsoft Excel.

3.8 Data Integration and Interpretation

The integration of data from different acquisition modes such as data-dependent acquisition (DDA) and data-independent acquisition (DIA) as well as from FAIMS-enhanced runs is essential for generating a comprehensive proteomic profile. After mass spectrometry analysis, raw data are processed using specialized software to identify peptides, quantify protein abundances, and detect any sequence variants or post-translational modifications. Data from DDA are analyzed to generate detailed spectra of individual peptides, while DIA provides more comprehensive coverage across the sample, capturing all ionized peptides in a mass range. By combining results from both modes, the depth of protein identification is enhanced, increasing the reliability of the findings. FAIMS data, in particular, allow for improved sensitivity and the detection of low-abundance proteins, especially in samples with limited input.

3.8.1 Data Processing and Protein Identification

1. Import raw MS/MS data into Proteome Discoverer, MaxQuant [14], Fragpipe [15], or any suitable proteomics software for protein identification. For DDA data, assign peptide spectra to protein sequences based on MS/MS fragmentation patterns. For DIA data, use spectral libraries or de novo sequencing tools to extract peptide information from complex MS/MS spectra.

2. Perform a database search using a reference protein database (e.g., SwissProt or a custom amyloid protein database) with the following parameters: Enzyme: Trypsin, allowing up to two missed cleavages. Fixed modification: Carbamidomethylation of cysteine. Variable modifications: Oxidation of methionine, deamidation of asparagine/glutamine. Peptide mass tolerance: 10 ppm. MS/MS fragment mass tolerance: 0.02 Da.
3. Filter the results to include only high-confidence protein identifications (peptide FDR $< 1\%$).

3.8.2 Quantification of Protein Abundance

1. For label-free quantification, use Proteome Discoverer or DIA-NN software to calculate protein abundances. DDA: Use spectral counting or MS1 precursor intensity to quantify protein levels. DIA: Quantify proteins based on MS2 fragment ion intensity or extracted ion chromatograms (XIC).
2. Normalize the quantified protein abundances across samples to account for variations in sample loading or instrument conditions.
3. Perform statistical analyses (e.g., t-test or ANOVA) to identify differentially abundant proteins between sample groups (*see* **Note 5**).

3.8.3 Sequence Variant and Post-Translational Modification (PTM) Analysis

1. For variant analysis, perform error-tolerant searches using Mascot or PEAKS de novo sequencing.
2. Identify potential amyloid protein variants (e.g., transthyretin V122I/L variant) in the MS/MS spectra.
3. Confirm variants by examining mass shifts and MS/MS fragmentation patterns.
4. Analyze post-translational modifications (PTMs) by identifying characteristic mass shifts (e.g., phosphorylation, glycosylation) in the MS/MS data.
5. Search for PTMs using specific tools in the software, such as PEAKS PTM analysis.

3.8.4 Data Interpretation

1. Compile all identified and quantified proteins, along with sequence variants and PTMs, into a comprehensive protein list.
2. Use gene ontology and pathway analysis tools (e.g., STRING, DAVID) to interpret biological processes related to the identified proteins.
3. Focus on processes related to amyloid fibril formation, immune response, or tissue remodeling.
4. Integrate the data into a final report that highlights significant findings, including the identified amyloid protein subtypes, their associated sequence variants, and any relevant biomarkers for disease classification.

4 Notes

1. Congo red-positive regions will appear red or salmon-pink under bright-field.
2. Typically, 60,000–180,000 square micrometers of tissue.
3. Notably, for our experiments in the reference publication [13], we utilized Thermo Fisher Scientific orbitrap-based technology (Orbitrap Fusion Lumos and Orbitrap Exploris 480). But other high-resolution mass spectrometers can be used.
4. Proteome raw files from orbitrap technology (Thermo Fisher Scientific) can be searched using Proteome Discoverer software (Thermo Fisher Scientific) using SEQUEST search engine. Alternatively, other free software are available as briefly discussed. Please refer to comprehensive proteomics reviews to explore in more detail the options based on your needs.
5. A detailed explanation of how to analyze proteomics datasets using a Microsoft Excel spreadsheet was recently published by Aguilan and colleagues [16]. This reference demonstrates to those who are not familiar with math and statistics the basic steps of –omics data processing.

Acknowledgments

The Sidoli lab gratefully acknowledges for funding the Hevolution Foundation (AFAR), the Einstein-Mount Sinai Diabetes center, and the NIH Office of the Director (S10OD030286).

Conflict of Interest Statement The authors declare no financial conflicts of interest.

References

1. Quock T, Yan T, Chang E et al (2018) Epidemiology of AL amyloidosis: a real-world study using US claims data. Blood Adv 2(10): 1046–1053. https://doi.org/10.1182/bloodadvances.2018016402
2. Benson M, Buxbaum J, Eisenberg D et al (2018) Amyloid nomenclature 2018: recommendations by the International Society of Amyloidosis (ISA) nomenclature committee. Amyloid 25(4):215–219. https://doi.org/10.1080/13506129.2018.1549825
3. Vrana J, Gamex J, Madden B et al (2009) Classification of amyloidosis by laser microdissection and mass spectrometry-based proteomic analysis in clinical biopsy specimens. Blood 114(24):4957–4959. https://doi.org/10.1182/blood-2009-07-230722
4. Dasari S, Theis J, Vrana K et al (2020) Amyloid typing by mass spectrometry in clinical practice: a comprehensive review of 16,175 samples. Mayo Clin Proc 95(9):1852–1864. https://doi.org/10.1016/j.mayocp.2020.06.029
5. Guan S, Talyor P, Han Z et al (2020) Data dependent-independent acquisition (DDIA) proteomics. J Proteome Res 19(8): 3230–3237. https://doi.org/10.1021/acs.jproteome.0c00186

6. Chapman J, Goodlett D, Masselon C (2014) Multiplexed and data-independent tandem mass spectrometry for global proteome profiling. Mass Spectrom Rev 33(6):452–470. https://doi.org/10.1002/mas.21400
7. Longuespee R, Alberts D, Pottier C et al (2016) A laser microdissection-based workflow for FFPE tissue microproteomics: important considerations for small sample processing. Methods 104:154–162. https://doi.org/10.1016/j.ymeth.2015.12.008
8. Venne K, Bonneil E, Eng K et al (2005) Improvement in peptide detection for proteomics analyses using NanoLC-MS and high-field asymmetry waveform ion mobility mass spectrometry. Anal Chem 77(7):2176–2186
9. Woo J, Clair G, Williams S et al (2022) Three-dimensional feature matching improves coverage for single-cell proteomics based on ion mobility filtering. Cell Syst 13(5):426–434 e424. https://doi.org/10.1016/j.cels.2022.02.003
10. Sweet S, Chain D, Yu W et al (2022) The addition of FAIMS increases targeted proteomics sensitivity from FFPE tumor biopsies. Sci Rep 12(1):13876. https://doi.org/10.1038/s41598-022-16358-1
11. Sethi S, Theis J, Leung N et al (2010) Mass spectrometry-based proteomic diagnosis of renal immunoglobulin heavy chain amyloidosis. Clin J Am Soc Nephrol 5(12):2180–2187. https://doi.org/10.2215/CJN.02890310
12. Hill M, Dasari S, Mollee P et al (2021) The clinical impact of proteomics in amyloid typing. Mayo Clin Proc 96(5):1122–1127. https://doi.org/10.1016/j.mayocp.2020.12.002
13. Aguilan J, Lim J, Racine-Brzostek S et al (2024) Effect of dynamic exclusion and the use of FAIMS, DIA and MALDI-mass spectrometry imaging with ion mobility on amyloid protein identification. Clin Proteomics 21(1):47. https://doi.org/10.1186/s12014-024-09500-w
14. Cox J, Mann M (2008) MaxQuant enables high peptide identification rates, individualized p.p.b.-range mass accuracies and proteome-wide protein quantification. Nat Biotechnol 26(12):1367–1372. https://doi.org/10.1038/nbt.1511
15. Yu F, Haynes SE, Nesvizhskii AI (2021) IonQuant enables accurate and sensitive label-free quantification with FDR-controlled match-between-runs. Mol Cell Proteomics 20:100077. https://doi.org/10.1016/j.mcpro.2021.100077
16. Aguilan JT, Kulej K, Sidoli S (2020) Guide for protein fold change and p-value calculation for non-experts in proteomics. Mol Omics 16(6):573–582. https://doi.org/10.1039/d0mo00087f

Chapter 11

Chromatin Immunoprecipitation Sequencing and RNA Sequencing from Laser-Microdissected Human Lung Cancer Cells

Emi Hanawa, Naomi Kaneko, Masayuki Noguchi, and Masafumi Muratani

Abstract

Genomic and epigenomic analyses are widely used in various biomedical research fields, with chromatin immunoprecipitation sequencing (ChIP-seq) representing one of the key techniques. The regulatory state of chromatin is crucial for understanding gene regulation mechanisms and the roles of transcription factors, and observing these states within native tissues using clinical specimens is particularly meaningful. One of the major advantages of ChIP-seq is its ability to comprehensively map histone modifications and transcription factor binding across the genome, especially those associated with specific functions. Additionally, it enables direct sequencing of genomic regions surrounding these modifications. Unlike techniques such as Assays for Transposase-Accessible Chromatin with high-throughput sequencing (ATAC-seq), which involve enzymatic digestion of nearby DNA, ChIP-seq preserves the surrounding DNA, allowing for the identification of nucleotide sequences adjacent to target proteins or histone modifications. This feature makes it especially effective for detecting single nucleotide polymorphisms (SNPs) and somatic mutations associated with allele-specific transcription factor binding or histone marks. This chapter describes a method for ChIP-seq analysis of clinical lung cancer tissue samples isolated by laser microdissection. The method is also compatible with RNA isolation and subsequent RNA-seq, enabling the acquisition of cancer cell-specific epigenomic profiles within heterogeneous tissue environments.

Key words Cancer, Clinical tissue sample, Pathology, Chromatin immunoprecipitation sequencing, Epigenome, Laser microdissection

1 Introduction

Extensive studies using cellular models and developmental processes in various organisms have revealed that distinct histone modifications are associated with specific regulatory states of genomic regions. For example, trimethylation of histone H3 at lysine 4 (H3K4me3) marks gene promoters, while acetylation at lysine 27 (H3K27ac) marks active promoters and enhancer regions [1–3]. Additionally, H3K4me1 is associated with enhancers and promoters, and H3K27me3 marks repressed developmental genes. By

Roberta Noberini (ed.), *Laser Capture Microdissection: Methods and Protocols*, Methods in Molecular Biology, vol. 3015, https://doi.org/10.1007/978-1-0716-5154-4_11,

analyzing combinations of multiple histone modifications, researchers can infer a wide range of regulatory states. Transcription factors, owing to their functional specificity, also contribute directly to the inference of gene regulatory networks. Against this background, ChIP-seq has become an increasingly important tool in medical research, including studies using clinical specimens such as cancer tissues.

ChIP-seq is also effective as a "capture sequencing" method, enabling efficient enrichment and analysis of DNA from regulatory regions of the genome. Allele-specific read counts can reveal allele-specific epigenetic regulation associated with SNPs and somatic mutations. This makes ChIP-seq a powerful approach, particularly when analyzing clinical samples that often exhibit high genomic variability. For example, in a study on gastric cancer by Muratani et al., histone modification changes between normal and cancer tissues were identified, and nearby SNPs and somatic mutations were further pinpointed, allowing for the prioritization of functionally relevant variants from a large pool of genomic alterations [4]. Similarly, a study by Quang et al. on lung cancer identified candidate genetic variants associated with epigenomic alterations [5]. Through peak calling of histone modifications, comparison with normal tissues, and filtering based on differences in read counts for each variant, it has been shown that functionally relevant variant candidates can be identified even from a small number—or even a single case—of clinical samples.

Despite its many advantages, performing ChIP-seq on clinical tissue specimens presents several technical challenges. One major issue is tissue heterogeneity. In cancer tissues composed of various cell types, the presence of normal epithelial cells, immune cells, and stromal cells can dilute the allelic frequency of somatic mutations specific to cancer cells. Similarly, epigenomic signals—particularly changes in histone modifications—can be masked by signals originating from other cell types. To address these issues, laser microdissection (LMD) is an effective technique for isolating specific cell populations [6]. Additionally, the quality of clinical tissue specimens often falls short of the optimal standards required for ChIP-seq. This chapter provides a detailed discussion of practical strategies for overcoming these limitations.

2 Materials

2.1 Clinical Tissue Sample Storage

1. Clinical tissue samples (approved by Institutional Review Board).
2. Cryovials.
3. Dry ice.

2.2 Tissue Fixation and Processing

1. Phosphate-buffered saline (PBS): 137 mM NaCl, 8.1 mM Na2HPO4, 2.68 mM KCl, 1.47 mM KH2PO4 (pH 7.4).
2. Formaldehyde solution (1%): Diluted from 37% in PBS before use.
3. Surgical blade.
4. OCT (optimal cutting temperature) compound.
5. OCT block mold.

2.3 Sectioning and Staining

1. Cryostat.
2. Glass slides.
3. LMD PEN Membrane Slide (Leica).
4. 1% Mayer's hematoxylin solution.

2.4 Laser Microdissection

1. LMD system: Leica LMD6000.
2. PCR tubes: 0.2 mL flat-cap tubes fitting to the LMD system.

2.5 RNA Isolation

1. Buffer PKD: RNeasy FFPE Kit (Qiagen).
2. Thermal cycler.
3. TRIzol-LS reagent (Invitrogen).
4. Snap-cap DNA LoBind 1.5 mL tubes.
5. Benchtop centrifuge.
6. Glycogen (10 mg/mL): Diluted from 20 mg/mL stock.
7. Isopropanol.
8. 80% ethanol: Prepared from 99.5% ethanol, diluted with distilled water.
9. Distilled water.
10. Bioanalyzer (Agilent Technologies).
11. RNA 6000 Pico Kit (Agilent technologies).

2.6 Buffers for ChIP

1. Sodium dodecyl sulfate (SDS) lysis buffer: 50 mM Tris–HCl (pH 8.0), 10 mM EDTA, 1% SDS.
2. Protease inhibitor: cOmplete™ ULTRA Tablets Mini, EDTA-free.
3. Dilution buffer: 10 mM Tris–HCl (pH 8.0), 140 mM NaCl, 1 mM EDTA, 1% Triton X-100, 0.01% SDS.

2.7 Antibody Beads

1. Protein G beads: Dynabeads™ Protein G, Invitrogen.
2. Magnetic stand.
3. H3K4me3 antibody: Millipore, Cat# 07-473.
4. H3K27ac antibody: Abcam, Cat# ab4729.
5. H3K27me3 antibody: Millipore, Cat# 07-449.

2.8 ChIP

1. Bioruptor sonicator (Diagenode).
2. Protein low-binding tubes.
3. TE buffer: 10 mM Tris–HCl (pH 8.0), 1 mM EDTA.
4. SDS elution buffer: 50 mM Tris–HCl (pH 7.5), 10 mM EDTA, 1% SDS.
5. Pronase stock solution: Pronase, dissolved in distilled water at 20 mg/ml and aliquot stored at −20 °C.
6. Phenol:chloroform:isoamyl alcohol (PCI), 25:24:1.

2.9 ChIP DNA Enrichment Monitoring and PCR Analysis

1. Human *ACTB* gene promoter primers:

   Forward: AAGCCGGCCTTGCACAT.

   Reverse: GCTATTCTCGCAGCTCACCAT.

   (Prepared as 5 pmol/μL stock by combining equal volumes of 100 pmol/μL forward and reverse primers)
2. Human *ACTB* gene outside primers:

   Forward: GAGCAGCTCCTGGATCACACT.

   Reverse: GCCAGTTAACTCTGGAAGCATGT
3. Fast SYBR™ Green Master Mix: Thermo Fisher Scientific, Cat# 4385616.

2.10 Sequencing Library Preparation

1. NEBNext Ultra II DNA Library Prep Kit for Illumina: New England Biolabs.
2. Index primers: NEBNext Multiplex Oligos for Illumina (Unique Dual), New England Biolabs.

3 Methods

3.1 Clinical Tissue Sample Storage

1. Surgical specimens are initially stored at 4 °C in the pathology laboratory for diagnostic purposes. After diagnostic evaluation, the remaining tissue is divided into small fragments of approximately 100 mg and placed into cryotubes for long-term storage. This step is usually performed at room temperature. To prevent tube rupture, the tissue is placed directly into a −80 °C freezer without snap-freezing in liquid nitrogen (*see* **Note 1**). It is not necessary to precisely weigh the tissue fragments; as a guideline, a volume appearing to be roughly 100 μL is sufficient.
2. Since previously frozen tissue specimens must not be subjected to repeated freeze–thaw cycles, it is recommended to prepare multiple aliquots using sufficiently small tissue fragments to avoid the need for refreezing.
3. Tissue samples can be stored for multiple years in a −80 °C freezer.

3.2 Tissue Fixation and Processing

1. Upon initial thawing, clinical tissue specimens obtained from the biobank are fixed in a solution of 1% formaldehyde in PBS for 10 minutes at room temperature. Fixation is performed in a standard 10 cm plastic disposable dish. If the tissue fragments are large and the fixative does not easily penetrate the interior, the tissue should be further dissected during the fixation process using a surgical blade to ensure effective crosslinking throughout the sample (*see* **Note 2**).
2. After the 10-minute fixation, the tissue fragments are transferred to PBS and incubated for an additional 10 minutes at room temperature. This step completes the fixation process and rinses away residual formaldehyde.
3. Tissue fragments are lifted using tweezers, and any excess PBS is carefully removed by gently blotting the tissue sample with a paper towel or Kimwipe. Each tissue piece is then placed into an OCT block mold prefilled with a small amount of Tissue-Tek OCT compound, taking care to avoid introducing air bubbles (*see* **Note 3**).
4. Additional OCT compound is added until the mold is completely filled. The mold is then placed on a layer of dry ice covered with aluminum foil to freeze the sample. Alternatively, this freezing step can be performed on the "quick freeze" stage of a cryostat (*see* **Note 4**).
5. Tissue samples embedded in OCT compound can be stored in a −80 °C freezer until sectioning.

3.3 Tissue Sectioning and Staining

1. Tissue sectioning is performed using a cryostat according to standard procedures. Serial sections are prepared for RNA analysis, ChIP assays, and histological evaluation. For RNA and ChIP analyses, 10 μm thick sections are mounted onto membrane slides compatible with LMD. For histological examination, 5 μm thick sections are mounted onto standard glass slides (Fig. 1; *see* **Note 5**).
2. Tissue sections mounted on slides are post-fixed in 1% formaldehyde in PBS at room temperature for 10 minutes.
3. After fixation, the slides are sequentially washed in PBS, then in distilled water, followed by hematoxylin staining for 4 minutes.
4. The stained slides are then rinsed with water, dehydrated through a graded ethanol series (70%, 95%, and 100%), and finally air-dried (*see* **Note 6**).

3.4 Laser Microdissection

1. Tissue dissection is performed using a Leica LMD system or any other LMD platform accessible to the researcher. If the investigator is inexperienced with performing ChIP assays using small amounts of input material, or if the target cell population represents only a small fraction of the tissue, it is

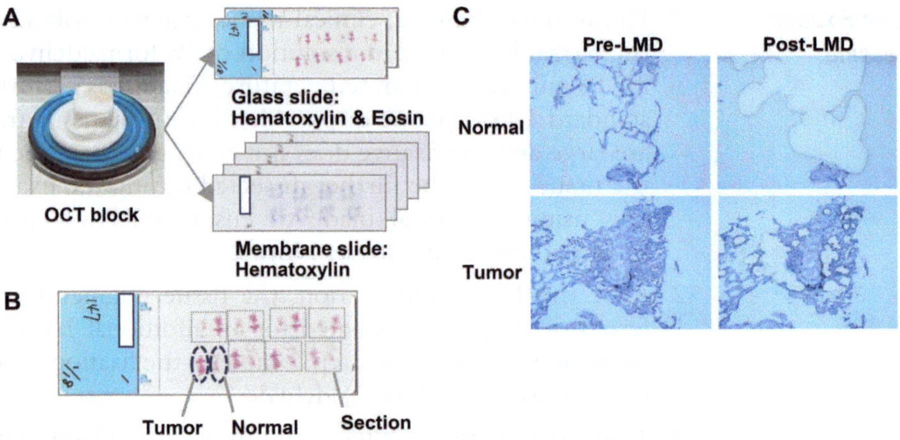

Fig. 1 Slide preparation and laser microdissection (LMD). (**a**) Serial sections are prepared and mounted onto glass and membrane slides. Glass slides are used for H&E staining to allow microscopic observation of tissue morphology and for archival purposes. Membrane slides, stained only with hematoxylin, are used for LMD. (**b**) Example of a H&E-stained glass slide. Normal and tumor tissue pairs are embedded within the same OCT block and sectioned. (**c**) Example of LMD, showing pre- and post-dissection images of normal and tumor tissue sections

advisable to collect dissected tissue fragments from multiple slides to ensure sufficient material for downstream processing (Fig. 2; *see* **Note** 7).

2. Tissue fragments isolated by LMD are collected in PCR tubes, stored at −80 °C, and processed at a later time for either ChIP assays or RNA purification, depending on the experimental schedule.

3.5 RNA Isolation (Optional)

1. A portion of the microdissected sample can be reserved in a separate collection tube specifically for RNA extraction (*see* **Note 8**). Add 160 μL of Buffer PKD containing proteinase K from the Qiagen RNeasy FFPE Kit, and incubate the sample at 56 °C for 15 minutes in a thermal cycler, followed by 80 °C for 10 minutes. After incubation, immediately cool the sample on ice.
2. Transfer the reverse-crosslinked sample in Buffer PKD to a fresh snap-cap DNA low-bind tube. Add TRIzol-LS reagent and distilled water to bring the final volume to 1 mL. At this stage, the sample can be stored at −80 °C for long-term preservation.
3. Proceed with the standard TRIzol protocol from this point onward (*see* **Note 9**). Add 200 μL of chloroform and vortex the tube vigorously for 30 seconds (*see* **Note 10**).
4. After proteinase K digestion and phase separation with TRIzol and chloroform, centrifuge the sample at 16,000 g for 15 minutes at 4 °C.

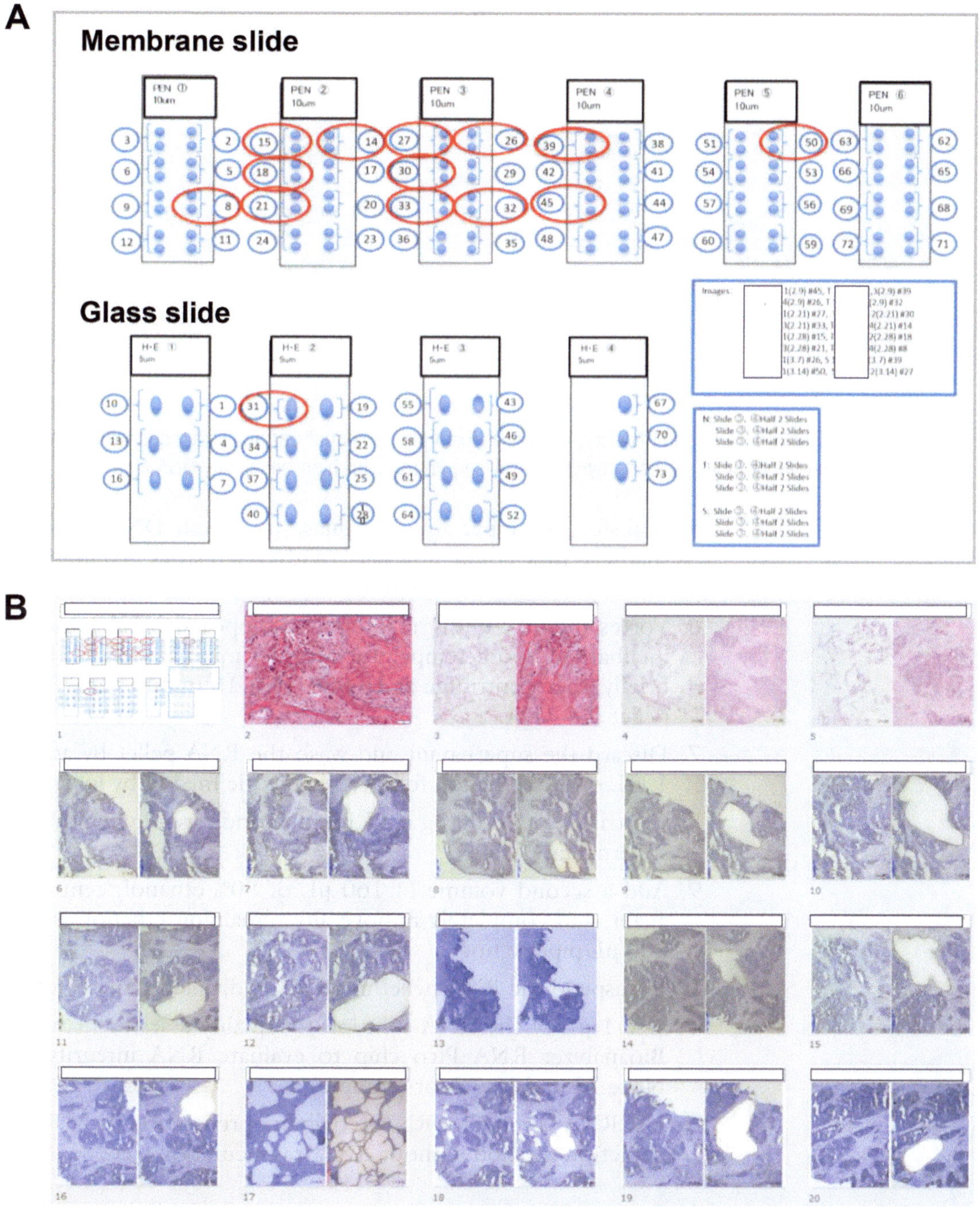

Fig. 2 Example of LMD records (**a**) Template showing slide layout and tissue organization. The order of serial sections and tissue images is annotated during sample processing. (**b**) Example of an LMD record and corresponding tissue image archive. Pre- and post-LMD images are retained for documentation purposes

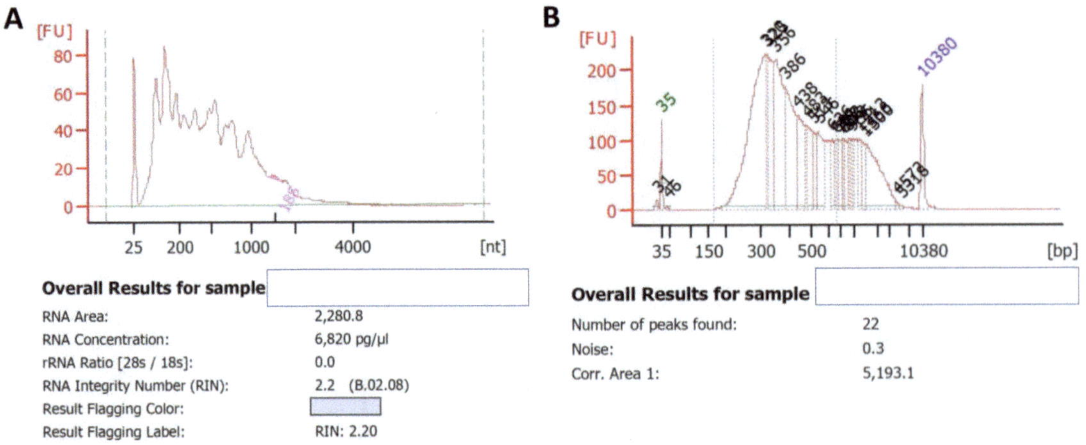

Fig. 3 Example of isolated RNA and RNA-seq library. (**a**) Example of bioanalyzer analysis result with RNA 6000 Pico chip. (**b**) Example of bioanalyzer analysis of RNA-seq library with high-sensitivity DNA chip

5. Transfer the upper aqueous phase to a fresh DNA low-bind snap-cap tube containing 1 μL of glycogen stock (10 mg/mL) as a carrier and add 500 μL of isopropanol.
6. Vortex the tube until the interface disappears completely, then incubate at room temperature for 10 minutes. Vortex again briefly, then centrifuge at 16,000 g for 15 minutes at 4 °C to pellet the RNA.
7. Discard the supernatant and wash the RNA pellet by adding 1 mL of 80% ethanol, followed by gentle mixing by inversion.
8. Centrifuge at 16,000 g for 5 minutes and carefully remove the supernatant.
9. Add a second volume of 160 μL of 80% ethanol, centrifuge again, and completely remove any remaining ethanol using a 200 μL pipette tip.
10. Resuspend the RNA pellet in 10 μL of distilled water.
11. Use 1 μL of the RNA solution for quality assessment with a Bioanalyzer RNA Pico chip to evaluate RNA integrity (*see* **Note 11** and Fig. 3 for example).
12. Use RNA samples for RNA-seq library preparation with rRNA-depletion method without RNA fragmentation step.

3.6 Buffer Preparation for ChIP Assay

1. Prepare SDS lysis buffer. Prepare a 10× inhibitor stock by dissolving one protease inhibitor mini tablet in 1 mL of SDS lysis buffer. Make 100 μL aliquots and store at −20 °C.
2. Combine nine parts SDS lysis buffer with one part 10× inhibitor stock. Use 100 μL per sonication sample.

3. Prepare dilution buffer. Prepare a 10× inhibitor stock by dissolving one protease inhibitor mini tablet in 1 mL of dilution buffer. Store at −20 °C.

3.7 Antibody Beads Preparation for ChIP Assay

1. Transfer 40 μL of Protein G Dynabeads per sample into a microfuge tube. Add 1 mL of dilution buffer (without inhibitors) for washing the beads and preparing the antibody–beads mixture.
2. Collect the beads on a magnetic stand, remove the buffer, and add 1 mL of dilution buffer.
3. Repeat **step 2** two more times to wash the beads. After the final wash, adjust the volume to 1 mL with dilution buffer.
4. In a fresh tube, combine 500 μL of dilution buffer (without inhibitors) with antibodies (1 μL per IP for H3K4me3, H3K27ac, and H3K27me3). Mix well by pipetting.
5. Add 500 μL of the bead suspension from step 3 (leaving the remaining half for preclearing) to the diluted antibodies to make a 1 mL antibody–beads mixture (*see* **Note 12**).
6. Rotate the tube in a cold room for 1 hour. Then collect the beads on a magnetic stand and remove the dilution buffer just before adding the chromatin solution for immunoprecipitation.

3.8 Tissue Lysis and Sonication for ChIP Assay

1. Bring the tissue pieces isolated by LMD from the freezer to room temperature.
2. Resuspend the tissue pieces in 100 μL of SDS lysis buffer (with protease inhibitor) by adding the buffer directly to the tube containing the isolated tissue. Use a small pipette tip to bring all tissue fragments to the bottom, ensuring they are immersed in the buffer. Incubate the sample on ice for 5 minutes.
3. Place the sample tubes in a sonicator water bath using a floating tube holder.
4. Sonicate the samples 12 times for 30 seconds each, with 30-second intervals, using the "High" setting (*see* **Note 13**).
5. The tissue should be dissociated. The lysis solution may appear purple due to the presence of dissociated hematoxylin-stained tissue.

3.9 Preclearing

1. After sonication, briefly spin the tube to bring all the liquid to the bottom.
2. Transfer the lysis buffer containing dissociated tissue to a fresh 1.5 mL Protein LoBind tube. Avoid collecting any LMD membrane pieces. Add 100 μL of inhibitor stock and 400 μL of washed beads from step 3.7 (without antibodies). Then add

400 μL of dilution buffer to bring the total volume to 1000 μL. Rotate at 4 °C for 1 hour.

3. Centrifuge at 16,000 g for 10 minutes to remove the preclearing beads and any insoluble cell debris or membrane fragments from the LMD slide.
4. Transfer the supernatant to a fresh 1.5 mL Protein LoBind tube. The volume of this chromatin solution should be approximately 900 μL. At this point, the SDS concentration is approximately 0.1%, and the NaCl concentration is approximately 140 mM. H3K4me3 ChIP can be performed under these conditions, but additional SDS and NaCl should be added as needed to optimize the immunoprecipitation conditions.
5. For H3K27ac and H3K27me3, add 10% SDS to adjust the final concentration to 0.2%.

3.10 Immunoprecipitation and Washing Beads

1. Remove the washing buffer from the antibody–beads complex tube prepared in step 3.7. Add 800 μL of chromatin solution and rotate the tube in a cold room overnight for immunoprecipitation. Store the remaining chromatin solution at −20 °C for use as the input control.
2. Collect the beads on a magnetic stand and wash them with 1 mL of dilution buffer. Repeat the wash two more times for a total of three washes.
3. After washing, remove the dilution buffer and add 500 μL of TE. Mix by pipetting to resuspend the beads and transfer the suspension into a new microfuge tube.
4. Add an additional 500 μL of TE to resuspend and recover remaining beads. Quickly spin down any residual beads using a benchtop mini centrifuge. Transfer and combine all the residual beads into the new tube.

3.11 Elution

1. Collect the beads on a magnetic stand and remove the TE. Add 100 μL of SDS elution buffer and resuspend the beads.
2. Incubate the beads at 68 °C for 10 minutes.
3. Recover the elution buffer using the magnetic stand and transfer it into new 0.6 mL PCR tubes.

3.12 Reverse Crosslinking

1. Prepare Pronase Mix by combining 100 μL of TE with 5 μL of Pronase stock solution (Proteinase K can also be used) for each reverse crosslinking reaction.
2. Add 100 μL of Pronase Mix to the eluted DNA.

3. For the input sample, combine 40 μL of chromatin solution (5% of total input), 60 μL of SDS elution buffer, and 100 μL of Pronase mix.
4. Incubate at 42 °C for 2 hours, then at 68 °C for 6 hours.

3.13 DNA Cleaning Up

1. Add 200 μL of PCI (phenol:chloroform:isoamyl alcohol) and extract. Centrifuge at 16,000 g for 5 minutes. Transfer the supernatant into a fresh 1.5 mL tube containing 15 μL of 5 M NaCl and 1 μL of glycogen (10 mg/mL stock).
2. Add 450 μL of ethanol. Mix and store at −20 °C for 4 hours or longer.
3. Centrifuge at 16,000 g for 30 minutes at 4 °C and remove the supernatant.
4. Wash with 180 μL of 70% ethanol. After spinning for 5 minutes, remove the supernatant completely.
5. Resuspend the DNA in 10 μL of water for whole-genome amplification or DNA library preparation.

3.14 Enrichment Monitoring Before DNA Library Preparation

1. Mix 1 μL of ChIP DNA with 9 μL of water to prepare a 1:10 dilution.
2. Perform real-time PCR reactions using 2 μL of the diluted DNA sample per primer pair (targeting the ACTB gene promoter and a negative control region outside the promoter).
3. The ACTB gene promoter is expected to be marked by H3K4me3 and H3K27ac in a wide range of cell types, in contrast to the negative control region amplified by the ACTB gene outside primers.

3.15 ChIP-Seq Library Preparation

1. Use 5 μL of undiluted ChIP DNA for library preparation with the NEBNext DNA Library Kit, following the manufacturer's protocol. Before the final PCR amplification step, use 1 μL of adaptor-ligated template for test amplification in the next step (*see* **Notes** 7 and **8**).
2. Perform real-time PCR analysis using the PCR cycling program indicated by the DNA library kit with the following reaction components:

1.0 μL	Adaptor-ligated DNA
5.0 μL	2x SYBR master mix
3.8 μL	H_2O
0.2 μL	Forward + reverse primer stock
10.0 μL	Total reaction volume

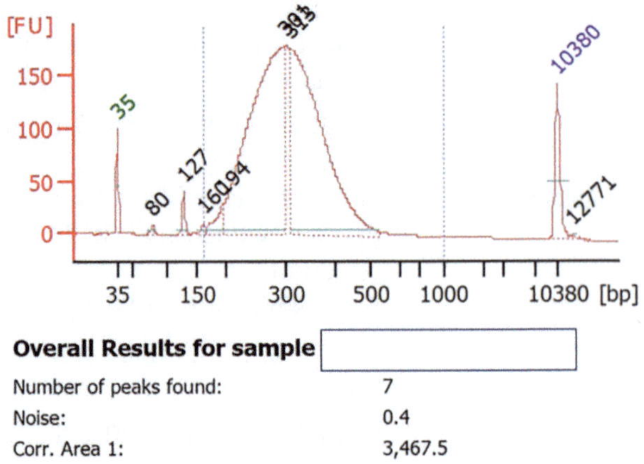

Fig. 4 Example of isolated RNA and RNA-seq library. Example of bioanalyzer analysis of DNA sequencing library with high-sensitivity DNA chip

3. This test amplification provides an estimate of the number of PCR cycles required in the final amplification step. Since the amount of DNA obtained from ChIP can vary, monitoring PCR efficiency is useful to ensure sufficient library yield for sequencing (Fig. 4).

4 Notes

1. Although protocols for performing ChIP assays on FFPE (formalin-fixed, paraffin-embedded) tissues have been reported, this chapter focuses on a method that utilizes fresh-frozen clinical tissue samples. In practice, tissue specimens are typically obtained from biobank facilities, where the preservation method may not always be under the researcher's control. Normally, freshly resected organs are kept at 4 °C for a few hours post-surgery, then dissected into smaller pieces, and stored directly at −80 °C. While it would be preferable to proceed immediately with the fixation step described in the next section prior to freezing, this is often impractical. In most clinical workflows, sample handling—including freezing—is carried out by biobank personnel, and it is generally difficult to perform fixation and embedding within the pathology laboratory. Alternatively, tissue can be frozen on aluminum foil placed on dry ice to achieve faster freezing. This is expected to yield better histological results but is often not part of standard biobank procedures.
2. The fixation described here is milder than that used for standard FFPE tissue preparation. Nevertheless, it is sufficient to preserve chromatin integrity and is well suited for ChIP assays.

The size of tissue fragments may vary depending on the clinical case, and fixation efficiency may differ between surface and deeper regions. Despite this variability, targets such as H3K4me3 and H3K27ac, which are generally robust in ChIP-seq assays, can be profiled with sufficient consistency to allow meaningful comparative analysis. It is important to note that once an unfixed tissue sample is thawed, chromatin structure and protein–DNA interactions can rapidly deteriorate. However, after this fixation step, the chromatin becomes stabilized and the sample can be handled at room temperature during subsequent processing.

3. In normal lung tissue, air bubbles are particularly prone to becoming trapped within the tissue interior. To restore proper tissue morphology, the sample can be placed into a syringe prefilled with an appropriate volume of OCT compound and subjected to vacuum. This procedure facilitates the penetration of OCT into the tissue, thereby improving structural integrity. While this step may be useful when histological detail is a priority, it is not necessary for ChIP assays. Even without this treatment, tissue sections of sufficient quality for laser microdissection (LMD) can generally be obtained.
4. In standard procedures for embedding fresh, unfixed tissue in OCT compound, rapid freezing is essential to preserve tissue morphology. This is typically achieved using metal containers floated in liquid nitrogen. However, in the present protocol, tissue fixation is performed prior to embedding for ChIP assays. As a result, strict temperature control during freezing is not required since tissue morphology and chromatin structure have already been stabilized by formaldehyde crosslinking.
5. Preparing serial sections from the same tissue block enables integrated analysis of tissue morphology, RNA expression, and chromatin profiles. This setup also facilitates downstream applications such as DNA isolation, mutation detection by Exome or amplicon DNA sequencing, and DNA methylation profiling, allowing for comprehensive multi-omics analysis. Although 20 μm thick sections can yield more biological material, they are more prone to detachment from the slide during staining. Since membrane slides used for LMD are relatively expensive, slides can be returned to room temperature after mounting a tissue fragment to allow sequential mounting of additional fragments. Once fully populated, the slide can proceed to staining. While OCT compound aids in cryosectioning, the tissue has already been fixed prior to sectioning, so handling at room temperature does not compromise sample quality.

6. Although the use of RNase-free reagents is not required during staining, general precautions against contamination are recommended—especially when slides will be used for downstream genomic analyses targeting low-frequency mutations. Ultralow-frequency mutations, particularly in cancer-related hotspots such as EGFR and KRAS, may be falsely detected due to tissue cross-contamination (e.g., from pathology lab equipment, staining baths, or wash solutions). Therefore, data interpretation should be carried out with caution.
7. Protocols for histone modification ChIP-seq using input equivalent to approximately 1000 cells have been established [7]. As a practical guideline, collect a visibly noticeable amount of tissue—typically seen as purple-stained fragments from hematoxylin staining accumulating in the collection tube. Visible fragments should contain at least 10,000 cells, which is sufficient to generate robust ChIP-seq peaks for H3K4me3 and H3K27ac when combined with low-input DNA library kits. For protocol optimization and parameter tuning, it is recommended to start with bulk tissue (large rectangular pieces from LMD membrane slides) before applying the method to microdissected samples with limited input.
8. Even among fresh-frozen tissue samples from the same biobank, not all specimens are suitable for ChIP assays. RNA quality can serve as a practical screening indicator. RNA extraction and quality assessment are less labor intensive than ChIP, and comparing RNA-seq and ChIP-seq data enables integrative epigenomic analysis. If RNA quality is too poor for RNA-seq, the sample is likely also unsuitable for ChIP-seq and may be excluded from further analysis.
9. Only the proteinase K digestion step from the RNeasy Kit is used; RNA purification after reverse crosslinking is carried out using the TRIzol reagent. This hybrid approach allows efficient recovery of small RNA quantities, which are finally resuspended in a minimal volume of distilled water.
10. The original TRIzol protocol recommends gentle mixing after adding chloroform. In this protocol, vigorous vortexing is recommended as DNA isolation is not being performed.
11. RNA quality is assessed using an Agilent Bioanalyzer with a Pico RNA chip. This provides information on RNA degradation. The extracted RNA can be used for RNA-seq library preparation with the TAKARA SMART kit. If the RNA integrity number (RIN) is 1, typically seen as a completely smooth curve, it indicates poor tissue preservation, and successful RNA-seq is unlikely. In such cases, the sample is also likely to be unsuitable for ChIP-seq. As the samples are fixed with formaldehyde, low RIN values (typically between 2 and 5) are expected.

12. It is important not to add antibody directly to the washed beads. To ensure even distribution, antibodies should first be diluted in a larger volume of dilution buffer, and then the beads should be added and mixed.
13. Chromatin fragmentation can be optimized using larger bulk tissue sections. However, optimal conditions may vary depending on cell type, fixation conditions, and tissue quality. Robust ChIP targets such as H3K4me3 and H3K27ac usually perform well under the same sonication conditions. Fragment size and potential sequencing bias can be evaluated during ChIP-seq data analysis.

Acknowledgments

Example images were derived from the study approved by the ethics committee of the University of Tsukuba Hospital (H29-052) and Institute of Medicine (1593-1). This work was supported by JSPS KAKENHI Grant Number JP15H05576 and 15H01343 to M.M.

References

1. Creyghton MP, Cheng AW, Welstead GG et al (2010) Histone H3K27ac separates active from poised enhancers and predicts developmental state. Proc Natl Acad Sci USA 107(50): 21931–21936
2. Rada-Iglesias A, Bajpai R, Swigut T et al (2011) A unique chromatin signature uncovers early developmental enhancers in humans. Nature 470(7333):279–283
3. Wang Z, Zang C, Rosenfeld JA et al (2008) Combinatorial patterns of histone acetylations and methylations in the human genome. Nat Genet 40(7):897–903
4. Muratani M, Deng N, Ooi WF et al (2014) Nanoscale chromatin profiling of gastric adenocarcinoma reveals cancer-associated cryptic promoters and somatically acquired regulatory elements. Nat Commun 5:4361
5. Ong Q, Sakashita S, Hanawa E et al (2021) Integrative RNA-Seq and H3 Trimethylation ChIP-Seq analysis of human lung cancer cells isolated by laser-microdissection. Cancers (Basel) 13(7)
6. Rabien A, Kristiansen G (2016) Tissue Microdissection. Methods Mol Biol 1381:39–52
7. Ng JH, Kumar V, Muratani M et al (2013) In vivo epigenomic profiling of germ cells reveals germ cell molecular signatures. Dev Cell 24(3): 324–333

Chapter 12

Laser-Capture Microdissection Followed by ATAC-Seq

Celia L. Hartmann, Joachim L. Schultze, Marc D. Beyer, and Caterina Carraro

Abstract

Laser-capture microdissection (LCM) enables the collection of regions of interest from target tissues, making it compatible with a number of spatially resolved downstream omics applications. Here, we provide a detailed protocol for LCM coupled to ATAC-seq (LCM-ATAC-seq) for the spatial analysis of chromatin landscape based on the method published by Carraro et al. (Cell Rep Methods 3:100598, 2023). LCM-ATAC-seq allows the efficient capture down to single nuclei and can be combined with the morphological staining of target cell populations to characterize chromatin states in fresh-frozen tissue samples.

Key words Laser-capture microdissection, Chromatin accessibility, Spatial omics, ATAC-seq, Morphological staining

1 Introduction

Laser-capture microdissection (LCM) allows for the isolation of cell subpopulations directly from tissue slices [1], thereby retaining the spatial information of the excised material [2]. Previous reports described the use of LCM coupled with a range of omics technologies to assess the genome, transcriptome, or proteome [3] of tissues with spatial resolution [4–8].

Spatially resolved ATAC-based epigenomics provides insight on context-dependent chromatin state regulation in tissues of interest. Different technologies, such as spatial ATAC-seq [9], sciMAP-ATAC [10], and solid-phase-capture spatial ATAC-seq [11], have been developed to investigate chromatin accessibility within a spatial context. However, these methods either lack spatial resolution or do not allow implementation of morphological marker-based selection of nuclei or regions of interest [1].

In this direction, we recently established a method for LCM coupled with ATAC-seq combined with immunostaining for specific cellular markers in order to select nuclei or regions of interest.

Roberta Noberini (ed.), *Laser Capture Microdissection: Methods and Protocols*, Methods in Molecular Biology, vol. 3015, https://doi.org/10.1007/978-1-0716-5154-4_12,

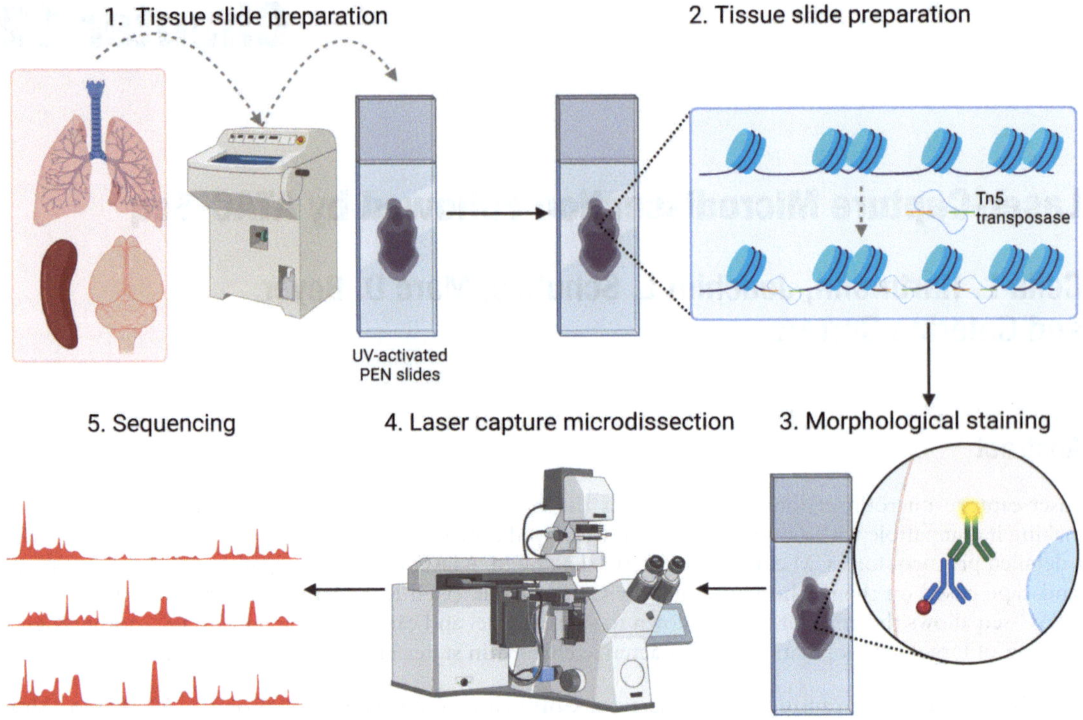

Fig. 1 Graphical representation of the main steps of the LCM-ATAC-seq protocol. Fresh-frozen tissues are first sectioned using a cryostat and then placed on a UV-treated PEN slide. In situ tagmentation with Tn5 transposase is followed by an optional morphological staining. After laser-capture microdissection, sample libraries are prepared and sequenced. (Created with BioRender)

LCM-ATAC-seq enables the targeted chromatin accessibility analysis of discrete either contiguous or scattered tissue populations via single-nuclei capture. We reported the efficient use of LCM-ATAC-seq on i) distinct fresh-frozen tissue types including mouse brain, lung, and spleen tissue as well as human lung tissue and at ii) distinct resolutions from minibulks of either big areas (~100–200 nuclei) or 50–100 single captured nuclei [1].

In this chapter, we provide a detailed protocol for LCM-ATAC-seq from slides preparation and in situ tagmentation, followed by optional morphological staining (Fig. 1). We then describe how to set up the microdissection and to prepare ATAC-seq libraries for sequencing [12]. Finally, we discuss a strategy for the preprocessing and quality control of the sequencing data. While this version of the protocol is optimized for LCM-ATAC-seq of human lung fresh-frozen tissues coupled with immunostaining of PU.1-positive cells, we further suggest how to adapt its use on different tissue types and target proteins.

2 Materials

In the following section, we provide a list of required reagents and solutions. Please note that solutions for fixation and staining are optimized for PU.1 immunostaining of human lung tissue.

2.1 Tissue Slide Preparation

1. Cryomold.
2. Optimal cutting temperature tissue embedding medium.
3. Isopentane.
4. Polyethylene naphthalate (PEN) slides (Carl Zeiss).
5. Cryostat and histology tools necessary for sectioning (brushes, scalpel).

2.2 In Situ Tagmentation

1. PAP pen (Vector Labs).
2. THS-seq buffer: 33 mM Tris-acetate pH 7.8, 66 mM potassium acetate, 10 mM magnesium acetate, 16% DMF (in nuclease-free water).
3. Transposition buffer: 0.01% digitonin in THS-seq buffer.
4. Tn5 enzyme preloaded with Illumina adapters (in-house or commercial; *see* **Note 1**).
5. Incubator/oven.
6. Humidity chamber.

2.3 Optional: PU.1 Morphological Staining of Lung Parenchyma Sections

1. Fixation solution: 1% paraformaldehyde (PFA) (*see* **Note 2**) in water (*see* **Note 3**).
2. PBST: 0.4% Triton X-100 in PBS.
3. Blocking buffer: 2% normal goat serum diluted 1:5 with PBST (*see* **Note 4**).
4. Primary antibody solution: Anti-PU.1 (E.388.3 monoclonal rabbit IgG anti-human PU.1, Thermo Fisher) diluted 1:100 with 1% bovine serum albumin (BSA, in PBS) (*see* **Note 5**).
5. Secondary antibody solution: Secondary antibody (goat anti-Rabbit IgG (H + L). Cross-Adsorbed Secondary Antibody, Alexa Fluor 488, Invitrogen) diluted 1:250 with 1% BSA in PBS (*see* **Note 6**).
6. 1 μM 4′,6-diamidino-2-phenylindole (DAPI) solution (in nuclease-free water).
7. KimTech tissues.

2.4 Laser-Capture Microdissection

1. PALM MicroBeam laser-capture microdissection system (Zeiss).
2. PCR tubes.

3. Reverse crosslinking buffer: 50 mM Tris–HCl pH 8, 50 mM NaCl, 20 μg/mL Proteinase K, 0.2% SDS in nuclease-free water.
4. Collection buffer (per sample): 2.5 μL of reverse crosslinking buffer and 2 μL of unique dual indexes mix (IDT for Illumina) diluted 1:5 in nuclease-free water.
5. CapMover.
6. SingleCapCollector 200 (RoboMover PALM equipment (Zeiss)).
7. 80% ethanol.
8. Centrifuge.
9. Thermal cycler.

2.5 Library Preparation and Sequencing

1. 10% Tween-20 in nuclease-free water.
2. NEBNext HiFi PCR Master Mix (New England Biolabs).
3. SeraMag Select magnetic beads.
4. TapeStation 4200 instrument.
5. HS D5000 tape.
6. Qubit HS dsDNA.

2.6 Software and R Packages

1. PALM RoboSoftware (Zeiss).
2. bcl2fastq2 (Illumina).
3. Trimmomatic [13].
4. Bowtie2 [14].
5. Picard [15].
6. deeptools [16].
7. samtools [17].
8. MACS2 [18].
9. MultiQC [19].
10. GenomicAlignments [20].
11. BRGenomics [21].
12. ATACseqQC [22].

3 Methods

As previously mentioned, the following protocol is optimized for LCM-ATAC-seq on fresh-frozen human lung tissue coupled with staining of PU.1-positive cells. Steps including tissue cutting parameters and morphological stainings need to be optimized for the tissue and protein target of interest. We provide notes where this is applicable.

3.1 Tissue Slide Preparation

1. Embed tissue specimen (*see* **Note 7**) into cryomolds filled with optimal cutting temperature (OCT) compound. Quickly place the cryomold with the tissue in precooled isopentane on dry ice and wait until the cryopreservation medium turns completely white. Store OCT-embedded tissue blocks at −80 °C.
2. Activate polyethylene naphthalate (PEN) slides (Carl Zeiss) in a UV crosslinker (UVP) with 254 nm wavelength light for 30 min with 200 mJ/cm^2 energy exposure.
3. For lung tissue, set the cryostat specimen and blade temperature to −18 °C and −20 °C, respectively (*see* **Note 8**). Cut lung tissue blocks into 10 μm thick sections with a cryostat and collect the sections on the UV-activated PEN slides (*see* **Note 9**). Keep PEN slides containing sections on dry ice before proceeding with the next steps (*see* **Note 10**).

3.2 In Situ Tagmentation

1. Thaw slides at room temperature for 1 min.
2. *Optional:* If opting for a morphological protein staining, proceed with *tissue fixation*. For PU.1 lung immunostaining, incubate slides in 1% PFA fixation solution for 10 minutes at room temperature (*see* **Note 11**). Please note that PFA fixation may affect the quality of the final chromatin profiles, as discussed in the original publication.
3. Wash slides with PBS for 5 minutes to completely remove OCT medium (as well as fixative, if applied) and wipe out excess PBS with a KimTech tissue. Be careful not to damage the PEN membrane.
4. Draw a hydrophobic contour on the slide around the tissue sections with a PAP pen (*see* **Note 12**). This will keep the solutions on top of the section during the incubation time, avoiding any spillover.
5. Dilute the Tn5 enzyme in the transposition buffer at the concentration suggested by the manufacturer. We diluted our in-house Tn5 stock 1:50 to a final concentration of 1.5 μM. This step may require a titration to optimize the enzyme concentration.
6. Apply 50 μL of Tn5 solution diluted in transposition buffer to the tissue section and incubate the slide for 1 h at 37 °C in a humidity chamber (*see* **Note 13**). Ensure that the entire tissue section is covered, adjusting reaction volume proportionally if required.
7. Wash slides three times with PBS. Proceed to Subheading 3.3 if a morphological staining is applied; otherwise, leave the last PBS layer on top of the section and proceed to Subheading 3.4 (laser-capture microdissection).

3.3 Optional: PU.1 Morphological Staining of Lung Parenchyma Sections

Morphological staining is an optional step of LCM-ATAC-seq which enables the targeted selection of specific regions or cell populations of interest. As mentioned in previous notes, this step needs to be optimized based on species, tissue type, and target antigen. We suggest increasing antibody concentrations compared to the vendor's instructions and reducing incubation times in order to minimize the processing time between thawing of the slide and microdissection, which can have an impact on the final quality of chromatin profiles. We report here an optimized protocol for the PU.1 nuclear staining of fresh-frozen human lung sections.

1. Incubate sections with 100 μL blocking buffer for 45 min (*see* **Note 14**) at room temperature.
2. Remove blocking buffer and incubate sections with 100 μL of primary antibody solution for 15 min at room temperature (*see* **Note 15**).
3. Wash tissue sections three times with PBST.

   Perform the following steps in the dark:

4. Incubate sections with 100 μL of secondary antibody solution for 15 min at room temperature. Make sure to choose a fluorescent dye compatible with the PALM system filters.
5. Wash sections three times with PBST.
6. Apply 100 μL DAPI for reference staining of nuclei and incubate for 5 min at room temperature.
7. Wash three times with PBS and keep the last layer on top of the section while proceeding to 3.4.

3.4 Laser-Capture Microdissection

1. Select regions or nuclei of interest (*optional*: based on morphological staining; Fig. 2) at the desired magnification using a laser-capture microscope (e.g., Zeiss PALM MicroBeam).
2. Carefully remove the excess of PBS from the top of the tissue section using a KimTech tissue.
3. Apply 80% ethanol to the tissue section and leave it to air-dry for no longer than 30 minutes (*see* **Note 16**).
4. Finely tune the microdissection parameters, that is, cutting and catapulting energy. If using the PALM RoboSoftware, use the *Calibration Wizard* function. Excessive energy can lead to tissue damage while insufficient energy may result in the incomplete dissection and/or catapulting of target tissue areas.
5. Fill PCR tube caps with 4.5 μL collection buffer and place into the PALM *SingleCap Collector*.
6. Cut tissue areas and collect the sample. Inspect the PCR collection cap using the *Cap Check* function to confirm the successful collection of dissected areas (*optional*).

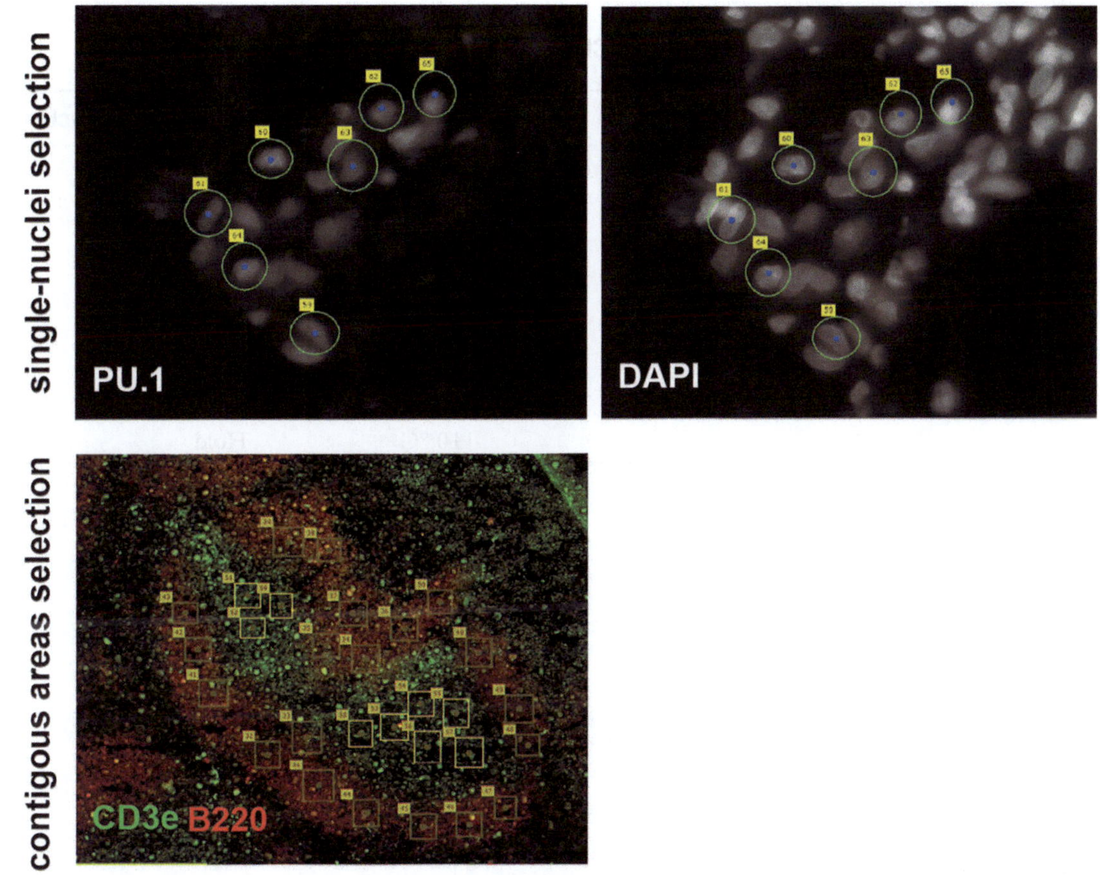

Fig. 2 Morphological staining-based selection strategies for LCM-ATAC-seq (*single-nuclei* or *continuous areas*). Representative image of a single-nuclei capture of human lung PU.1-positive nuclei (PALM microscope, 40× magnification, *upper* panels) and of a contiguous area selection of B- and T-cell mouse spleen zones based on B220 and CD3e immunofluorescence staining (PALM microscope, 5× magnification, *lower* panel)

7. Centrifuge the PCR tube containing the microdissected sample at 1000 g for 30 s to collect everything at the bottom. Keep tubes on ice if multiple samples are collected.
8. Place the PCR tube containing the microdissected sample on a thermocycler at 65 °C with lid temperature set to 100 °C overnight (for PFA-fixed material) to reverse crosslink the Tn5 enzyme. For unfixed material, 1 h incubation is sufficient to release the Tn5.

3.5 Library Preparation and Sequencing

1. Add 5 μL of 10% Tween 20 to the samples and mix by pipetting.
2. Add 10 μL NEBNext HiFi PCR Master Mix to the samples and mix by pipetting.

Table 1
PCR program for LCM-ATAC-seq library preparation

	Temperature	Time	Cycles
Tagmentation end repair	72 °C	10 min	1
Initial denaturation	98 °C	5 min	1
Denaturation	98 °C	10 s	21
Primer annealing	63 °C	30 s	
Extension	72 °C	20 s	
Final extension	72 °C	5 min	1
	10 °C	Hold	–

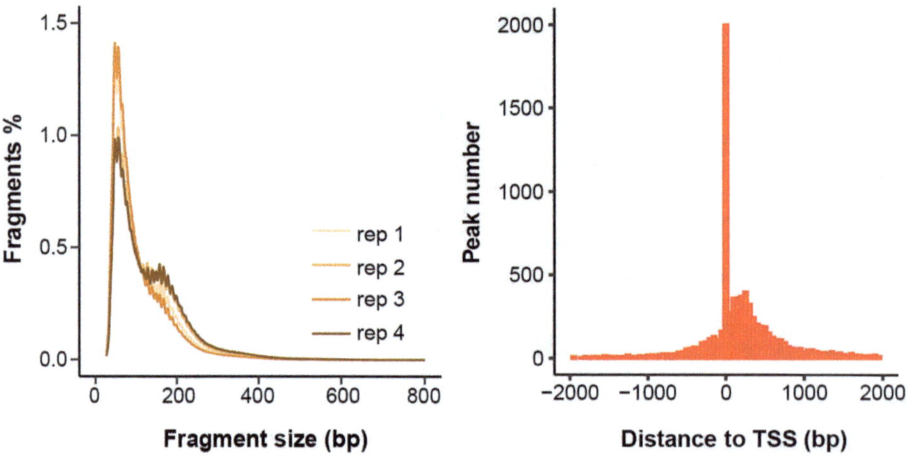

Fig. 3 Representative QC metrics of LCM-ATAC-seq libraries. Classical fragments-size distribution (FSD) with mono- and di-nucleosome pattern (*left*) and TSS enrichment profile of LCM-ATAC-seq reads (*right*)

3. Place PCR tubes on thermocycler and run the program reported in Table 1.
4. Briefly spin down the tubes and clean up the PCR products using 20 μL SeraMag Select magnetic beads, following the manufacturer's instructions.
5. Check the library size distribution using the HS D5000 assay on a TapeStation 4200 instrument. A typical ATAC-seq footprinting distribution shows generally up to a di-nucleosomes peaks pattern, as discussed in the original publication (Fig. 3 *right* panel). This may vary based on the quality of the analyzed tissue block and the specific tissue preprocessing protocol.
6. Check the library concentration using Qubit HS dsDNA assay.

7. Sequence the libraries on a NovaSeq6000 instrument using SP or S1 v1.5 chemistry (Illumina) paired-end 2 × 50 cycles.
8. PCR products can be stored long-term at −20 °C.

3.6 Preprocessing and Quality Control of the LCM-ATAC-Seq Dataset

As a final step, we report here an example of the suggested preprocessing pipeline for the bioinformatic evaluation of the LCM-ATAC-seq library quality. For QC evaluation, classical ATAC-seq metrics should be investigated as defined below and further discussed in the original publication (*see* **Note 17**).

1. Demultiplex the sequencing data using bcl2fastq2 (Illumina).
2. Perform adapter trimming using Trimmomatic [13].
3. Align the sequencing reads against a reference genome using Bowtie2 [14] (*see* **Note 18**).
4. Remove duplicated reads using the Picard *dedup* function [15].
5. Correct the Tn5-induced offset using the deeptools *alignmentSieve* function [16].
6. Perform sorting and indexing of bam files using samtools [17].
7. Utilize MACS2 for peak calling [18].
8. Generate alignment statistics via the MultiQC tool [19].
9. Quantify sequencing reads in these unified peak regions in R using the *summarizeOverlaps* function in the GenomicAlignments [20] package.
10. Prefilter raw counts to exclude low-count peaks (<10 reads).
11. Check the following ATAC-seq parameters for proper quality evaluation of the obtained libraries: alignment quality control, fragment size distribution (FSD, Fig. 3 *left* panel), transcription start site enrichment scores (TSS-Ess; Fig. 3 *right* panel) (BRGenomics [21], ATACseqQC [22]) log10(unique reads), fraction of total reads falling in peaks (FRiP) (defined as reads in peaks/(reads in gaps+reads in peaks)), and percentage of mitochondrial reads.

4 Notes

1. Store Tn5 enzyme stock at −20 °C and keep on ice while in use.
2. Concentration and type of fixation depend on species, tissue type, target antigen, and selected staining antibody (e.g., for B220 and CD3e staining of mouse spleen sections, use 2% PFA solution in water, as reported in the original publication).
3. Prepare the fixation solution fresh before each use and properly dispose of chemicals following local regulations.

4. Adjust blocking buffer composition depending on the species, tissue type, target antigen, and selected staining antibody (e.g., for B220 and CD3e staining of mouse spleen sections, use 5% normal goat serum diluted 1:2 in PBST, as reported in the original publication).
5. Optimize primary antibody solution depending on species, tissue type, target antigen, and selected staining antibody (e.g., for B220 and CD3e staining of mouse spleen sections, use 1:25 dilution in PBST with 5% goat serum 2.5% BSA, as reported in the original publication).
6. Optimize secondary antibody solution depending on species, tissue type, target antigen, and selected staining antibody (e.g., for B220 and CD3e staining of mouse spleen sections, use 1:25 secondary antibody in PBST 2.5% BSA, as reported in the original publication).
7. This protocol has been optimized for fresh-frozen tissue. Working with formalin-fixed paraffin-embedded (FFPE) tissue specimens may require additional optimization of the protocol, with a potential impact on ATAC-seq profiles based on the lower starting quality of chromatin substrate.
8. Temperatures may be adjusted based on tissue type.
9. Adjust tissue thickness based on species and tissue type.
10. If not processed immediately, store PEN slides with sections of interest in sealed containers with desiccant at −80 °C for no longer than 1 month.
11. This step is optional and only necessary when additional antibody staining is required. Time of incubation and composition of the fixation solution may vary depending on tissue type, target antigen, and selected staining antibody.
12. Ensure that the barrier does not touch the tissue sections. Make sure to leave a little space between the sections and the barrier to be able to add the reagents onto the tissue sections. Let the barrier dry completely before proceeding to the next step.
13. Place water or water-soaked tissues into the humidity chamber before use to ensure that tissue sections do not dry out.
14. Optimize blocking solution and incubation time based on species, tissue type, and target antigen.
15. Optimize antibody solutions and incubation time based on species, tissue type, target antigen, and chosen primary antibody.
16. Longer incubation time may lead to tissue shrinkage and affect the quality of ATAC-seq data.

17. It is important to note that the overall quality of the obtained libraries depends on a number of factors including the i) starting quality of tissues, ii) the microdissected areas and total number of collected nuclei per sample, iii) the presence of a PFA fixation step, and iv) the length of the overall morphological staining protocol.
18. Overall, QC metrics should still be in the range of other spatial ATAC-seq methodologies as reported in the original publication.
19. Select an appropriate reference genome. For human and mouse tissues, we used hg38 human reference and mm10 mouse reference genome, respectively.

Acknowledgments

We thank particularly Dr. Adem Saglam, the Light Microscope Facility at DZNE Bonn, and the other co-authors of the original publication for their valuable contribution to the establishment of the methodology reported in this chapter.

References

1. Carraro C, Bonaguro L, Srinivasa R et al (2023) Chromatin accessibility profiling of targeted cell populations with laser capture microdissection coupled to ATAC-seq. Cell Rep Methods 3:100598. https://doi.org/10.1016/j.crmeth.2023.100598
2. Guo W, Hu Y, Qian J et al (2023) Laser capture microdissection for biomedical research: towards high-throughput, multi-omics, and single-cell resolution. J Genet Genomics 50: 641–651. https://doi.org/10.1016/j.jgg.2023.07.011
3. Datta S, Malhotra L, Dickerson R et al (2015) Laser capture microdissection: Big data from small samples. Histol Histopathol 30:1255–1269. 10.14670/HH-11-622
4. Moulédous L, Hunt S, Harcourt R et al (2003) Proteomic analysis of immunostained, laser-capture microdissected brain samples. Electrophoresis 24:296–302. https://doi.org/10.1002/elps.200390026
5. Schillebeeckx M, Schrade A, Löbs A-K et al (2013) Laser capture microdissection-reduced representation bisulfite sequencing (LCM-RRBS) maps changes in DNA methylation associated with gonadectomy-induced adrenocortical neoplasia in the mouse. Nucleic Acids Res 41:e116. https://doi.org/10.1093/nar/gkt230
6. Nichterwitz S, Chen G, Aguila Benitez J et al (2016) Laser capture microscopy coupled with Smart-seq2 for precise spatial transcriptomic profiling. Nat Commun 7:12139. https://doi.org/10.1038/ncomms12139
7. Herrera JA, Mallikarjun V, Rosini S et al (2020) Laser capture microdissection coupled mass spectrometry (LCM-MS) for spatially resolved analysis of formalin-fixed and stained human lung tissues. Clin Proteomics 17:24. https://doi.org/10.1186/s12014-020-09287-6
8. Shen S, Li J, Huo S et al (2021) Parallel, high-quality proteomic and targeted metabolomic quantification using laser capture microdissected tissues. Anal Chem 93:8711–8718. https://doi.org/10.1021/acs.analchem.1c01026
9. Deng Y, Bartosovic M, Ma S et al (2022) Spatial profiling of chromatin accessibility in mouse and human tissues. Nature 609:375–383. https://doi.org/10.1038/s41586-022-05094-1
10. Thornton CA, Mulqueen RM, Torkenczy KA et al (2021) Spatially mapped single-cell chromatin accessibility. Nat Commun 12:1274. https://doi.org/10.1038/s41467-021-21515-7
11. Llorens-Bobadilla E, Zamboni M, Marklund M et al (2023) Solid-phase capture and

profiling of open chromatin by spatial ATAC. Nat Biotechnol 41:1085–1088. https://doi.org/10.1038/s41587-022-01603-9

12. Buenrostro JD, Wu B, Chang HY, Greenleaf WJ (2015) ATAC-seq: a method for assaying chromatin accessibility genome-wide. Curr Protoc Mol Biol 109:21.29.1–21.29.9. https://doi.org/10.1002/0471142727.mb2129s109
13. Bolger AM, Lohse M, Usadel B (2014) Trimmomatic: a flexible trimmer for Illumina sequence data. Bioinformatics 30:2114–2120. https://doi.org/10.1093/bioinformatics/btu170
14. Langmead B, Salzberg SL (2012) Fast gapped-read alignment with Bowtie 2. Nat Methods 9:357–359. https://doi.org/10.1038/nmeth.1923
15. Picard Tools – By Broad Institute. https://broadinstitute.github.io/picard/. Accessed 13 May 2025
16. Ramírez F, Dündar F, Diehl S et al (2014) deepTools: a flexible platform for exploring deep-sequencing data. Nucleic Acids Res 42:W187–W191. https://doi.org/10.1093/nar/gku365
17. Li H, Handsaker B, Wysoker A et al (2009) The sequence alignment/map format and SAMtools. Bioinformatics 25:2078–2079. https://doi.org/10.1093/bioinformatics/btp352
18. Zhang Y, Liu T, Meyer CA et al (2008) Model-based analysis of ChIP-Seq (MACS). Genome Biol 9:R137. https://doi.org/10.1186/gb-2008-9-9-r137
19. Ewels P, Magnusson M, Lundin S, Käller M (2016) MultiQC: summarize analysis results for multiple tools and samples in a single report. Bioinformatics 32:3047–3048. https://doi.org/10.1093/bioinformatics/btw354
20. Lawrence M, Huber W, Pagès H et al (2013) Software for computing and annotating genomic ranges. PLoS Comput Biol 9:e1003118. https://doi.org/10.1371/journal.pcbi.1003118
21. DeBerardine M (2023) BRGenomics for analyzing high-resolution genomics data in R. Bioinformatics 39. https://doi.org/10.1093/bioinformatics/btad331
22. Ou J, Liu H, Yu J et al (2018) ATACseqQC: a Bioconductor package for post-alignment quality assessment of ATAC-seq data. BMC Genomics 19:169. https://doi.org/10.1186/s12864-018-4559-3

Chapter 13

Laser Capture Microdissection for Quantitative Shotgun Lipidomics

Silvia Radrezza, HongKee Moon, Ivey Sebastian, Gayathri Nadar, Lena Hersemann, and Andrej Shevchenko

Abstract

In tissues or organs, lipids are not distributed uniformly. While staining with dyes or antibodies locates histological features at cryosections of tissues, it does not elucidate their molecular composition or infer their biological function. Here, we describe how to determine the molar abundance of hundreds of molecular species of major lipid classes, including membrane and neutral lipids, in the lipidome of spatially confined areas of tissue cryosections using laser capture microdissection coupled with shotgun mass spectrometry.

Key words Shotgun lipidomics, Spatial lipidomics, Laser capture microdissection, Mass spectrometry

1 Introduction

Staining with dyes, such as hematoxylin/eosin (H&E), or with specific antibodies is widely used to visualize histological features on the sections of tissues and image them using light or electron microscopy [1–3]. However, staining provides no evidence on the molecular composition of their proteome, lipidome, or metabolome.

Matrix-assisted laser desorption/ionization (MALDI) imaging maps a selection of individual biomolecules, most commonly lipids or metabolites, directly at tissue sections with a spatial resolution down to a few micrometers [4]. Despite being sensitive and specific, MALDI imaging relies on the difference in relative intensities of molecular ions across neighboring areas, which is not necessarily reflecting molar quantities of different biomolecules. The ionization capacity strongly differs between lipid classes, which may bias compositional inference for neutral lipids (e.g., diacylglycerides or triacylglycerides) and polar glycerophospholipids (e.g., phosphatidylcholines or phosphatidylethanolamines). Furthermore, the

Roberta Noberini (ed.), *Laser Capture Microdissection: Methods and Protocols*, Methods in Molecular Biology, vol. 3015, https://doi.org/10.1007/978-1-0716-5154-4_13,

abundance of peaks acquired from seemingly identical spatially defined features cannot be summed across broader cryosection areas, which limits the quantification capacity and lipidome coverage.

Alternatively, tissue material from several visualized features can be recovered using laser capture microdissection (LCM), combined, and analyzed together [5]. In this approach, one of the two successively cut slices is stained to image the regions of interest (ROIs) using bright-field or fluorescence microscopy. However, staining could affect the lipidome composition. Therefore, the image is then matched to the second unstained slice by in-house developed software and ROIs are excised by LCM Lipids are extracted from the collected material and analyzed by shotgun mass spectrometry. Material from morphologically identical features can be pooled together to enhance the lipidome coverage.

This approach enables fast, comprehensive, and accurate molar quantification of major lipid classes and species, allowing direct comparison of the full lipid composition of spatially specific regions isolated from different samples. In this way, LCM and shotgun mass spectrometry contribute to the field of spatial lipidomics [6] by reliable molar quantification of 150–200 species from 15 major lipid classes using as little as ~0.5 mm^2 of 20 micrometers-thick tissue slices. This chapter explains practical implementation of the method.

2 Materials

2.1 Cryodissection

1. CryoStar NX70 (Thermo Fisher Scientific).
2. Snap frozen tissue, stored on dry ice until its dissection.
3. Microtome blade: Carbon steel, C35 (Feather).
4. Glass slides (Superfrost™ Plus Adhesion Microscope).
5. Sample holder.
6. O.C.T. (optimal cutting temperature) compound or Neg50.
7. Ice-cold acetonitrile (ACN) (ca. 4–6 °C) in 50 mL Falcon tube.

2.2 LCM and ROI Image Transfer

1. Vacuum desiccator.
2. PALM MicroBeam System (Carl Zeiss Microscopy) or another suitable microscope.
3. 0.2 mL plastic tubes (Sigma-Aldrich).
4. Isopropanol (IPA), LC grade.
5. ROI image transfer software: https://doi.org/10.5281/zenodo.15586320.

2.3 Lipid Extraction

All solvents should be of LC grade or higher quality. Synthetic lipid standards were purchased from Avanti Polar Lipids (Alabaster, AL) and stored at −20 °C until use.

1. SpeedVac or a vacuum desiccator.
2. Tips, 10–1000 μL (Thermo Fisher Scientific).
3. 0.5 mL tubes (Thermo Fisher Scientific).
4. Extraction solution: Methyl-*tert*-butyl ether (MTBE)/methanol (MeOH), 10:3 v/v.
5. Internal standards (IS) mixture: for each sample, an aliquot of 100 μL of the internal standard mix in MTBE/MeOH 10:3 v/v containing 25 pmol of Chol-d7, 32 pmol of CE-d7 16:0, 18 pmol of TG-d5 50:0, 9 pmol of DG-d7 33:1, 24 pmol of PC 25:0, 14 pmol of LPC 13:0, 8 pmol of PE 25:0, 3 pmol of LPE 13:0, 3.5 pmol of PI 25:0, 2 pmol of PG 25:0, 3 pmol of Cer 30:1:2, 4 pmol of PA-d7 33:1, 4 pmol of SM 30:1:2, and 2 pmol of PS 25:0. Stock solutions of individual standards in MeOH were mixed to reach the concentrations listed above. *See* **Note 1** for acronyms.
6. Thermomixer placed in a cold room at 4 °C.
7. Centrifuge with 0.5 mL tubes rotor.
8. 2 mL amber glass vials (Supelco) and compatible solid caps PTFE (Sigma-Aldrich).
9. SprayMix solution: 4:2:1 isopropanol/methanol/chloroform (IPA/MeOH/$CHCl_3$) containing 7.5 mM ammonium formate in MeOH.

2.4 Shotgun Mass Spectrometry

1. Twin.tec 96-wells PCR plate, skirted, colorless, 150 μL (Eppendorf SE).
2. Robotic nanoflow ion source TriVersa NanoMate (Advion Biosciences).
3. Nanoelectrospray chips with 4.1 μm orifice diameter (Advion Biosciences).
4. Q Exactive hybrid tandem mass spectrometer (Thermo Fisher Scientific).

2.5 Data Analysis

1. PeakStrainer software [7].
2. LipidXplorer software [8].
3. Molecular fragmentation query language (MFQL) files used as lipid class-specific queries by LipidXplorer. For guideline examples, see https://lifs-tools.org/wiki/index.php/LipidXplorer_MFQL.

3 Methods

A protocol overview is shown in Fig. 1.

Typically, two cryo-slices are sequentially cut: one is used for staining, while another one is used for microdissecting the ROIs. Slices should look similar for better spatial matching accuracy. After staining and imaging, select eight to ten reproducible landmarks (e.g., tissue borders, vascular structures, or other distinct features) as reference points on the stained slice. For accurate matching, the reference points must be visible on both stained and unstained slices. Input reference points and image directories into the software and check the quality of matching. If mismatches occur (e.g., a points color or position does not align), manually correct reference points pairing using the "mapping" tool. Once images are aligned and matched, they are used by LCM instrument for isolating ROIs from the unstained slice. Once the tissue material is collected, known quantities of lipid standards are spiked into the tube and lipids are recovered by two-phase MTBE / methanol extraction.

3.1 Sample Preparation

1. Mount the snap frozen tissue in the sample holder with one to two drops of OCT or Neg50 and let it freeze. For each sample, obtain two consecutive 15–20 μm thick cryosections at −25 °C (blade)/ −13 °C (holder) and place them onto separate glass slides (*see* **Notes 2** and **3**).

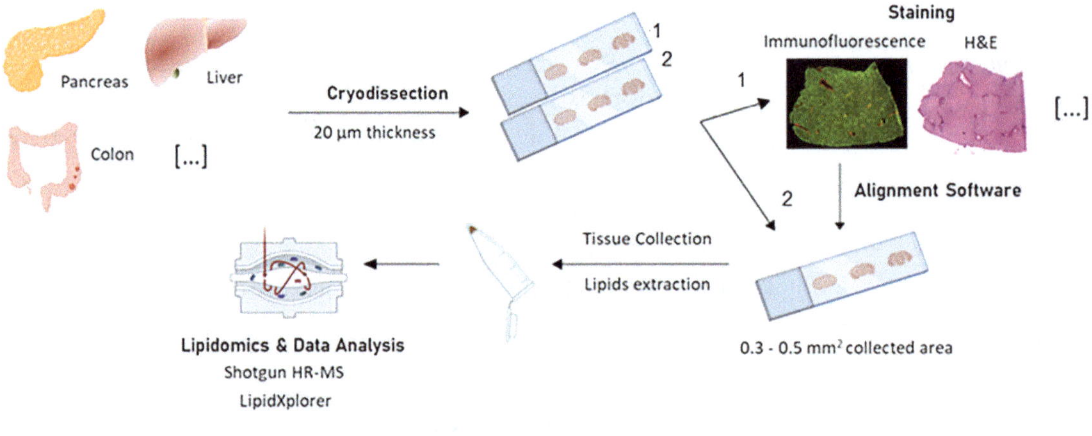

Fig. 1 Workflow for analyzing sliced tissues by shotgun LCM. Two 15–20 μm thick sequentially sliced cryosections are placed on individual glass slides. One section (1) is used for staining. Areas of interest are matched to specific locations at the unstained section (2) using a dedicated software, laser ablated, and catapulted into the Eppendorf tubes. Lipids are then extracted and quantified by shotgun analysis. (Figure created with BioRender.com)

2. Submerge the slide into ice-cold ACN for 1–2 seconds and let it dry for a few minutes at room temperature. Do this only for the slides that will not be stained later and keep them at −80 °C until further use (*see* **Notes 5** and **6**).
3. Stain only one of the two slides. The staining method (immunostaining, H&E, etc.) depends on the microscope and research purpose and is compatible with the workflow for as long as it produces a clear image of ROIs.

3.2 LCM Acquisition and ROI Image Transfer Software

1. Image the stained and unstained sections on the PALM MicroBeam System using the AxioCam ICc camera with the following settings: 5× magnification; excitation, 385 nm/470 nm at 50%/100%; reflector, 62 HE; excitation filter at Free; and exposure time at Auto (*see* **Note 4**).
2. Scan both stained and unstained sections and save them in.tiff format.
3. Select the ROIs in the stained section at 20x magnification and export them in.txt format.
4. Mark eight to ten identifiable common points in the stained and unstained sections (Fig. 2) to use as "reference points" and export them in.txt format (*see* **Notes 7–9**).
5. Open the "Info point" program and upload the scans to obtain x/y stage coordinates. This program comes with the PALM Robo software.
6. Open the Jupyter Notebook for ROI transfer software (https://doi.org/10.5281/zenodo.15586320) and run the sections in sequential order as outlined below (*see* **Note 10**).
7. For every pair of slices, update the following parameters in the first section of the ROI transfer software:
   - "X/Y stage position"= Coordinates obtained from the "Info point" software
   - "tiles" = image folders directory with.jpg and.zni image files included in.img folder
   - "initial file" = directory for the stained section (with folder and name of picture)
   - "initial points file"=.txt file with the reference points coordinates of the stained section
   - "rotated file" = directory for the unstained section
   - "rotated points file" =.txt file with the reference points coordinates of the unstained section
   - "elements file" = .txt of the ROIs selected in the stained slide.

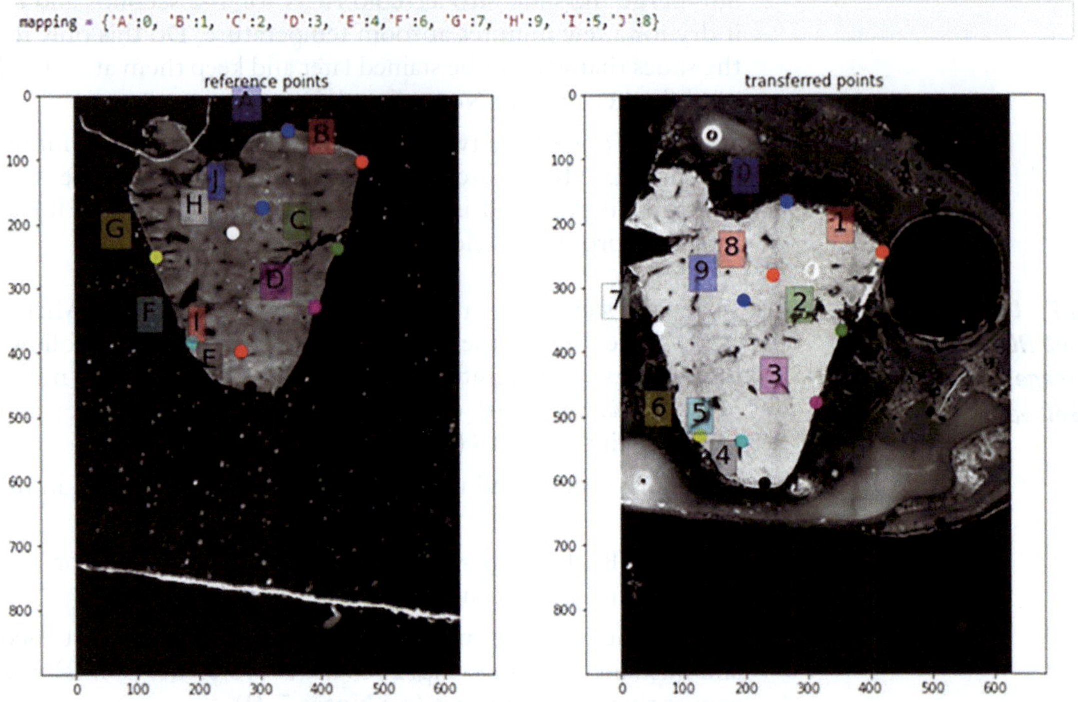

Fig. 2 The left side panel displays ten reference points (**a**–**h**) selected on the stained tissue section. The right panel shows these points transferred to the corresponding locations on the unstained section (labeled 1–9). The accuracy of reference points transfer is verified by comparing both the spatial positions and color matching of each point between the two panels. The color assignment does not guarantee correct spatial transfer; for instance, point C corresponds accurately to point 2, but for points H and 7, the transferred position (7) is incorrect. In such cases, the correct pairs (e.g., H–9) should be assigned manually using the mapping function in the software (the panel placed above the images)

8. Verify the concordance between reference points, adjusting the sequence if needed. Some reference points can be removed to allow better transfer. An example of reference points sequence is shown in Fig. 2.
9. Once the reference points are correctly matched between the stained and unstained sections, run the software to overlay the images. An example of overlay sections is shown in Fig. 3.
10. After the section are overlaid, the ROIs will be transferred onto the image of the unstained section. The coordinates for these transferred ROIs will be exported as an element file .txt.
11. This output.txt file can be imported into PALM Robo software, enabling automated cutting of ROIs based on the element list in the unstained slice.

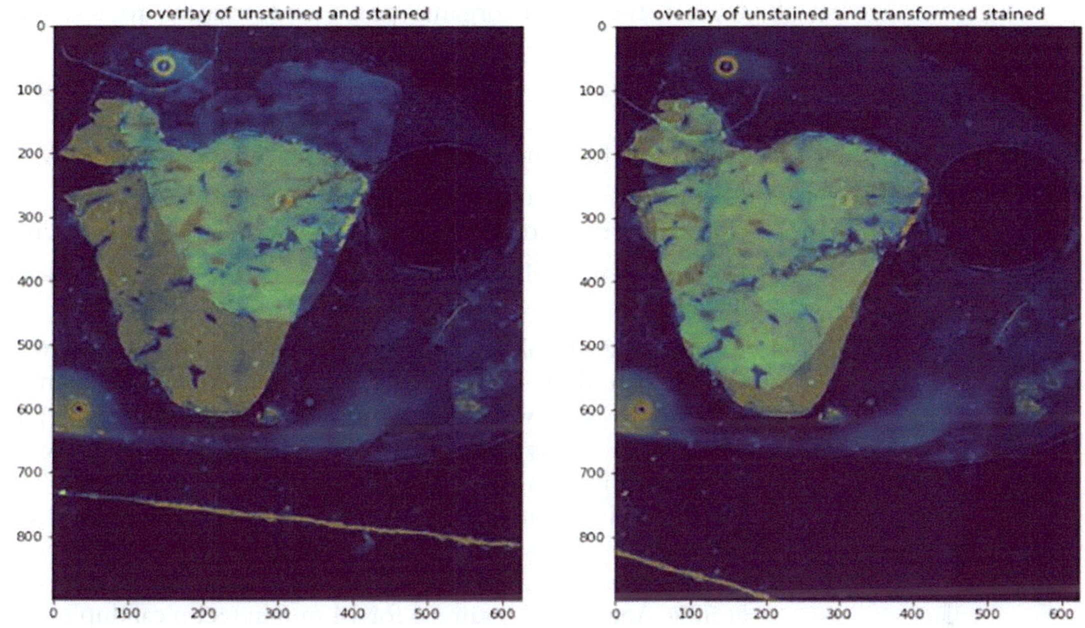

Fig. 3 Once the reference points are correctly matched, the two slices (unstained and stained) should be overlaid. In this example, the unstained slice appears slightly bigger (darker green). However, other structural features match. ROIs should only be selected within the overlapping area

3.3 Tissue Collection

1. Fill the 0.2 mL tube cap with 20 μL of isopropanol.
2. For LCM under automatic laser pressure catapulting (AutoLPC), we used the following settings: 20× magnification; energy: 5; focus: 7 (both in arbitrary units); 12 μm distance of AutoLPC shots; AutoLPC shots, horizontal (*see* **Note 4**).
3. Once the tissue is collected, close the tube, spin down, then rinse the cap with additional 20 μL of isopropanol, and spin it down (*see* **Notes 11–13**).
4. Store the collected material at −80 °C until further use.

3.4 Lipids Extraction

1. Equilibrate the samples at room temperature and dry them under vacuum.
2. Equilibrate the IS (internal standards) mixture at room temperature (*see* **Notes 14** and **15**).
3. Add 100 μL of IS mix in MTBE/MeOH 10:3 v/v to each sample (*see* **Notes 16** and **17**).
4. Shake for 1 h using a thermomixer (4 °C, 850 rpm).
5. Add 20 μL of H_2O.
6. Shake for 15 min, 4 °C at 850 rpm.
7. Centrifuge for 15 min, 4 °C at 13,400 rpm.

8. Collect the upper organic phase and transfer to a 2 mL amber vial.
9. Store the vials at −20 °C until further use.

3.5 Mass Spectrometry Analysis

1. Evaporate the sample in a desiccator or SpeedVac.
2. Reconstitute each sample with 100 μL of SprayMix if the collected area is around 0.5 mm^2. If less, 70 μL are sufficient.
3. Load a 96-well plate with 15–20 μL of the sample for each injection.
4. Analyze the samples through a Q Exactive equipped with a robotic nanoflow ion source TriVersa NanoMate using nanoelectrospray chips with a spraying nozzle diameter of 4.1 μm. The ion source is controlled by the Chipsoft 8.3.1 software (Advion Biosciences). The ionization voltage should be +0.96 kV in positive and −0.96 kV in negative mode, the backpressure 1.25 psi in both modes. The required temperature of the ion transfer capillary is 200 °C; the S-lens rf level is at 50%. Analyze each sample for 14 min in technical duplicates. Fourier transform (FT)-MS spectra are acquired within the range of m/z 300–1000 from 0 to 0.2 min in positive and within the range of m/z 300–1000 from 7.7 to 7.9 min in negative mode, with a mass resolution of 140.000 at m/z 200, automated gain control (AGC) of 3×10^6, and maximal injection time of 3000 ms. Parallel reaction monitoring in positive (0.2–1.7 min) needs to be conducted for the cholesterol acquisition, with precursor isolation window of 0.8 Th, nCE 12.5%, AGC 5×10^4, and maximum injection time of 3000 ms. Targeted selective ion monitoring (t-SIM) spectra in positive (1.7–7.5 min) and negative (7.9–14 min) modes are acquired with resolution of 140.000 at m/z=200 with AGC of 5×10^4, maximum injection time of 650 ms, isolation window of 20 Th, and scan range of m/z 300–1000. The inclusion list of masses targeted in t-SIM analyses should start at m/z 305 both in positive and negative mode, and other masses need to be computed by adding 10 Th increment (i.e., m/z 355, 365, 375, etc.) up to m/z 1005 [9] (*see* **Notes 18** and **19**).

3.6 Data Analysis

1. Preprocess the.raw files by PeakStrainer software [7]. The software employs repetition rate filtering to remove the background noise without compromising lipid class coverage.
2. Import.MZXML files obtained from PeakStrainer into LipidXplorer [8] for lipid identification. MFQL queries are compiled for each lipid class (PC, PC O-, LPC, LPC O-, PE, PE O-, LPE, PI, PG, PA, PS, SM, TG, DG, Cer, Chol, and CE). Glycerophospholipids (PC, PC O-, LPC, LPC O-, PE, PE

O-, LPE, PI, LPI, PG, PA, LPA, PS), SMs and DGs are analyzed in negative mode; TGs, Cer, and sterol lipids (Chol, CE) are analyzed in positive mode. Lipids are identified by matching accurately determined intact masses (mass accuracy better than 5 ppm). The identified lipids together with the abundance of their peaks are exported as a.csv file into the user's specified directory.

3. Quantify lipids by comparing isotopically corrected abundances of their molecular ions with the abundances of internal standards of the same lipid class.
4. Once quantified, we strongly recommend filtering the results based on the following criteria: do not consider lipids only detectable the blank sample; do not consider lipids if their concentration in the sample is 5 times lower than in the blank; consider lipids only if present in at least 50% of samples in the same batch. All these rules can be automated through an R or Python script to make the analyses faster.

4 Notes

1. Acronyms: PC: phosphatidylcholine; PC O-: ether-phosphatidylcholine; LPC: lyso-phosphatidylcholine; LPC O-: ether lyso-phosphatidylcholine; PE: phosphatidylethanolamine; PE O-: ether-phosphatidylethanolamine; LPE: lyso-phosphatidylethanolamine; PI: phosphatidylinositol; PG: phosphatidylglycerol; PA: phosphatidic acid; PS: phosphatidylserine; SM: sphingomyelin; TG: triacylglycerol; DG: diacylglycerol; Cer: ceramide; Chol: cholesterol; CE: cholesterol ester.
2. Leave glass slides at room temperature; chilled slides impact the adhesion. If collecting multiple sections on a single slide, keep the slide inside the cryostat between collections. If needed, warm up the target area by pressing with a finger on the reverse side of the glass slide.
3. It is important to obtain successive, reproducible and thin slices. Avoid exceeding 20–30 μm thickness as some features (e.g., bile canaliculi) may be positioned horizontally.
4. Ice-cold acetonitrile treatment for 1–2 sec preserves the lipid profile by rapidly inhibiting lipases. To prevent overexposure (which can cause tissue darkening or "burnt" appearance), adjust the solvent and immersion time based on tissue type and experimental goals. For example, other solvents like ice-cold MeOH for metabolomics or ice-cold EtOH for human pancreas can be considered. Change solvents after collecting one or two batches.

5. Label the glass slices with a pencil instead of a marker. Ink may dissolve and smeare by coming in contact with the solvent.
6. The temperature for the blade and holder and the LCM settings are provided as example. The parameters may need to be adjusted based on the tissue, slice thickness, external temperature/humidity, or the type of microscope (for LCM settings).
7. Having tissue slices with similar shape and position on the glass is critical for ROI image transfer and accurate LCM isolation.
8. Choose eight to ten easily recognizable reference points to have a quick visual check during transfer; edit or remove nonmatching reference points if possible.
9. If scanned slices show no good contrast between tissue and background, switch to the dark mode. This way, the tissue slice appears white on a black background.
10. Errors in the ROI transfer software can result from incorrect file path formatting (e.g., using "/" instead of "\").
11. Don't air-dry the slices for extended period of time. Over-dried tissue can be lost during LCM leading to unreliable material collection. We suggest processing slices within 30 minutes of exposure to room temperature.
12. For 20 μm thick slices analyzed by a high-resolution shotgun mass spectrometer, the optimal area to collect is around 0.5 mm^2. We recommend avoiding collecting areas smaller than 0.15 mm^2 to have a reliable signal quality.
13. Acquire images of tissue slices before and after the microdissection to verify the accuracy and quality of material collection.
14. For lipid analyses, ensure that the tips are suitable for organic solvents. We strongly recommend "Thermo Fisher Scientific" branded tips.
15. Prepare the internal standard mixture 1 day in advance and let it equilibrate at room temperature overnight.
16. Always include at least one blank sample containing only solvents and the internal standards mixture to monitor contamination and background signal.
17. The lipid composition varies between the samples. The internal standard mixture described above is optimized for mouse liver and collected areas of ca 0.5 mm^2. Adjust the composition of the internal standard mixture, if required.
18. Before mass spectrometry measurements, ensure the instrument is calibrated and well performing. Due to the limited sample material, use low reconstitution volumes and plan for two to three injections per sample.
19. The instrument settings described here were adjusted to our our specific equipment and research objections settings.

Acknowledgments

We thank Dr. Britta Schroth-Diez of the Light Microscopy Facility at the MPI-CBG for technical support with microscopy and Dr. Oskar Knittelfelder for his valuable contribution to software development.

References

1. Harms PW, Frankel TL, Moutafi M et al (2023) Multiplex immunohistochemistry and immunofluorescence: a practical update for pathologists. Mod Pathol 36(7):100197
2. Mebratie DY, Dagnaw GG (2024) Review of immunohistochemistry techniques: applications, current status, and future perspectives. Semin Diagn Pathol 41(3):154–160
3. Chan JK (2014) The wonderful colors of the hematoxylin-eosin stain in diagnostic surgical pathology. Int J Surg Pathol 22(1):12–32
4. Shevchenko A, Simons K (2010) Lipidomics: coming to grips with lipid diversity. Nat Rev Mol Cell Biol 11(8):593–598
5. Chan YH, Pathmasiri KC, Pierre-Jacques D et al (2024) Gel-assisted mass spectrometry imaging enables sub-micrometer spatial lipidomics. Nat Commun 15(1):5036
6. Varga-Zsíros V, Migh E, Marton A et al (2023) Development of a laser microdissection-coupled quantitative shotgun Lipidomic method to uncover spatial heterogeneity. Cells 12(3):428
7. Schuhmann K, Thomas H, Ackerman JM et al (2017) Intensity-independent noise filtering in FT MS and FT MS/MS spectra for shotgun Lipidomics A. Anal Chem 89:7046–7052
8. Herzog R, Schwudke D, Schuhmann K et al (2011) A novel informatics concept for high-throughput shotgun lipidomics based on the molecular fragmentation query language. Genome Biol 12(1):R8
9. Knittelfelder O, Traikov S, Vvedenskaya O et al (2018) Shotgun Lipidomics combined with laser capture microdissection: a tool to analyze histological zones in Cryosections of tissues. Anal Chem 90(16):9868–9878

Chapter 14

Artificial Intelligence-Driven Laser Capture Microdissection

Tommaso Gasparello, Matthew Pearson, and Alex von Kriegsheim

Abstract

Laser capture microdissection enables precise isolation of regions of interest (ROIs) from tissue sections, allowing researchers to investigate tissue heterogeneity with spatial resolution. This technique has been integrated with various omics approaches, including transcriptomics and proteomics, to enhance sensitivity and expand coverage beyond traditional bulk analysis methods. Among these, mass spectrometry (MS)-based proteomics has emerged as a particularly powerful tool by enabling the reliable identification of even low-abundance proteins and facilitating the discovery of novel disease biomarkers.

Recently, both laser capture microdissection and MS-based proteomics have been significantly shaped by the rapid advancement and widespread adoption of artificial intelligence (AI). In fact, deep learning, a subfield of AI, has been used to improve bioimage analysis for ROI selection in laser capture microdissection workflows. Concurrently, AI-based methods are increasingly being used to manage and interpret high-dimensional data generated by MS-based proteomics, further enhancing analytical depth and reproducibility.

Despite this progress, current workflows often lack scalability and flexibility, limiting their broader application in clinical and translational settings. To address these challenges, we have developed an AI-driven laser capture microdissection protocol optimized for hematoxylin and eosin (H&E)-stained slides derived from formalin-fixed paraffin-embedded (FFPE) patient samples. The widespread availability of FFPE tissues and the simplicity of H&E staining make this approach both accessible and adaptable for investigating diverse disease contexts, with the potential to advance spatial proteomics at scale.

Key words Laser capture microdissection, Mass spectrometry-based proteomics, Artificial intelligence, Formalin-fixed paraffin-embedded, Hematoxylin and eosin, Spatial proteomics, Deep visual proteomics

1 Introduction

Laser capture microdissection (LCM) has emerged as an efficient method for isolating regions of interest (ROIs) from heterogeneous tissues, hence allowing to preserve spatial context during sample preparation before conducting downstream expression analyses [1].

Recently, this technology has been coupled with multiple omics methodologies, including mass spectrometry (MS)-based

Roberta Noberini (ed.), *Laser Capture Microdissection: Methods and Protocols*, Methods in Molecular Biology, vol. 3015, https://doi.org/10.1007/978-1-0716-5154-4_14,

proteomics. The main attractions of MS-based proteomics are its specificity of detection and quantification, its sensitivity, and its unbiasedness (i.e., the fact that is not targeted against a specific protein) [2]. When applied after LCM, this method allows to achieve proteome-wide coverage and good detection of low-abundance proteins, previously a major limitation of protein research when compared to DNA and RNA studies (due to the lack of amplification methods) [3].

Today, MS-based proteomics experiments can generate large-scale, multidimensional datasets that are extremely challenging to analyze with classical statistical processes [2]. In this context, artificial intelligence (AI) is emerging as a powerful tool to conduct statistical analysis and deal with much larger and more complex datasets [2]. AI has not only been applied to assist with convoluted bioinformatics pipelines, but it has also been used in another field that is relevant for LCM. In fact, recently, AI-driven bioimage analysis has allowed to perform automated image segmentation and single-cell isolation, which can be used to guide the collection of specific ROIs from heterogeneous tissue sections when applied in conjunction with LCM workflows [4].

Recent advances in the fields of AI, LCM, and proteomics led to the development of a new technique known as deep visual proteomics (DVP, which is part of the broader field of spatial proteomics) [4]. This integrates proteome-wide data generation with advanced imaging techniques and aims to retain the spatial information of tissues by allowing to dissect individual cells.

Besides the conceptual appeal of spatial proteomics (and in particular of deep visual proteomics), one of its main advantages is that it can be readily applied to archived biobank tissues (i.e., formalin-fixed paraffin-embedded (FFPE) samples) [4, 5]. In fact, although oligonucleotide-based analysis of these tissues has often proven impossible due to the need for fixation and embedding in an inert matrix [6, 7], FFPE samples have been shown to be well suited to MS-based proteomic analysis [8]. Since FFPE is the most common method used to preserve human tissues for clinical diagnosis (with an estimate of at least 500 million cancer tissues stored to date) [9] and samples can be easily stored for many years, there is currently an abundance of material that could be used to conduct spatial proteomics experiments on large cohorts of patient-derived samples.

When dealing with FFPE samples, the staining process is particularly important to generate high-resolution microscopy images used to guide LCM. Recently, many groups have focused on developing immunofluorescence-based spatial proteomics pipelines [4, 5]. This allows to use specific biomarkers to reliably identify cells of interest and facilitate nuclear and cellular segmentation by clearly demarcating cellular boundaries. The main advantages of this approach are its reproducibility (i.e., cells are selected based on the objective expression of certain biomarkers) and the possibility to combine it with individual cell segmentation for downstream ultrasensitive MS-based analysis (when integrated into the deep visual proteomics framework) [2]. This approach, however, requires the use of previously known biomarkers to identify specific cells (or subcellular structures) of interest within a tissue and often entails lengthy optimization processes to fine-tune the staining and to enable reliable recognition of particular cell types. In addition to this, IF protocols are limited by their lack of flexibility as the markers used need to be constantly adapted to different tissues and cellular entities.

Due to these limitations, our group developed a pipeline to perform H&E-based spatial proteomics. This pipeline applies AI to bioimage analysis processes guiding LCM and to the differential expression analysis of MS-generated proteomics data. As discussed in the following sections, this method maximizes the quality of high-resolution microscopy images by using a xylene-based mounting medium (the gold standard for image clarity in most histopathology laboratories around the globe). It also includes bioimage analysis performed with an intuitive and open-access software, QuPath [10], which does not require any programming experience and is suited to both single-cell and population analysis.

2 Materials

2.1 Reagents

1. Triethylammonium bicarbonate (TEAB).
2. n-Dodecyl-beta-maltoside (DDM).
3. Acetonitrile (ACN) LC/MS grade.
4. Trifluoroacetic acid (TFA) HPLC grade.
5. 20 ng/μL endoproteinase Lys-C stock solution.
6. 20 ng/μL proteomics-grade modified trypsin stock solution.
7. Isopropanol LC/MS grade.
8. Tris(2-carboxyethyl)phosphine hydrochloride.
9. 2-Chloroacetamide (CAA).
10. Formic acid (FA) LC-MS grade.
11. Clear Vue Mountant XYL.
12. Harris hematoxylin, acidified.
13. Eosin Y aqueous.
14. Phosphate-buffered saline (PBS).
15. Vectabond.
16. Lysis buffer: 0.1% DDM buffer containing 5 mM TCEP, 20 mM CAA, 100 mM TEAB (pH 8.5) in milliQ water (can be stored in −80 °C). CAA and TCEP are toxic. Pipette in a fume hood with gloves. Make fresh before use.

17. Digestion buffer: 10% ACN, 0.1 M TEAB (pH 8.5), 2 ng/μL endoproteinase Lys-C, 2 ng/μL trypsin in milliQ water (freshly prepared). ACN is toxic. Pipette in a fume hood with gloves.

2.2 Equipment

1. Micro cover glass 24 × 50 mm.
2. PEN membrane glass slides (4 μm thickness) (Leica Microsystems).
3. Eppendorf 384 LoBind Plate.
4. CL-1000 Ultraviolet crosslinker UVP.
5. Leica RM2235 microtome.
6. Autostainer.
7. NanoZoomer XR.
8. Leica Laser Microdissection Microscope LMD7 (Leica Microsystems).
9. Humidity sensor.
10. Adaptors (as illustrated below).
11. Zero-Stat 3.
12. Centrifuge Mega star 4.0.
13. Speedvac.
14. Thermocycler.
15. Evosep Evotips.
16. Bruker timsTOF SCP (Trapped Ion Mobility/Time of Flight Single-Cell Proteomics).
17. Evosep one liquid chromatography separation device.
18. Ionoptiks Aurora Elite column 5 × 75 C18.
19. Zeiss Axioscan 7.

2.3 Software

1. Hamamatsu NanoZoomer manufacturer software.
2. QuPath version 0.6.0, including InstanSeg extension (*see* **Note 1**).
3. QuPath to LMD Streamlit app.
4. Leica LMD software (8.5 version).
5. Leica Las X.
6. Proteoscape.
7. DIA-NN 2.0.
8. Perseus.
9. Max Quant.
10. AlphaPepStat.
11. R.

12. py_diAID (Automated Isolation Design for Data-Independent Acquisition), a python package to support diaPASEF scan mode for mass spectrometry.

3 Methods

3.1 Tissue Cutting Using a Microtome for Formalin-Fixed Paraffin-Embedded (FFPE Tissue)

1. Pretreat the slides for 1 h at full power in a UV crosslinker.

   Optional: Coat the slides with Vectabond to improve tissue adherence to the slide.
2. Place the tissue blocks on the ice cube for approximately 1 h before use (store FFPE tissue blocks at room temperature).
3. Place the tissue block in the microtome holder and attach the microtome blade (remember to always place the knife guard over the microtome blade).
4. Cut tissue sections (5–10 μm thick based on cell types of interest).
5. Mount the sections on PEN membrane slides by slowly lifting the membrane side of the slide from underneath the floating tissue section.
6. Repeat as needed, periodically clean the blade from excess paraffin.
7. Dry the slides at 37 °C overnight.

3.2 Tissue Deparaffinization and Hematoxylin–Eosin (H&E) Staining

1. Deparaffinize FFPE samples by heating them in an oven at 60 ° C for 10 min and cooling them down for 5 min before proceeding to the next step.
2. Then wash the samples three times in xylene for 3 min, twice in 100% ethanol for 2 min, once in 80% ethanol for 2 min, once in 50% ethanol for 2 min, and once in milliQ water for 2 min.
3. H&E staining: Incubate slides in Mayer's hematoxylin for 1 min, rinse in water for 2 min, then in Scott's Tap Water Substitute (STWS) for 30 s followed by MilliQ water for 1 min, stain with eosin for 1 min, add Neoclear mounting media (a xylene substitute; *see* **Note 2**), apply a coverslip, and allow it to solidify for 1 h at room temperature.

3.3 High-Resolution Microscopy/Whole Slide

1. Scan the slide with NanoZoomer with the following settings: 40× magnification, automatic detection of scanning area, automatic detection of five focus points.

   Alternatively, it is possible to use other whole-slide imaging systems, such as Axioscan or the LAS X software (integrated software of the Leica LCM microscope; *see* **Note 3**) [5].

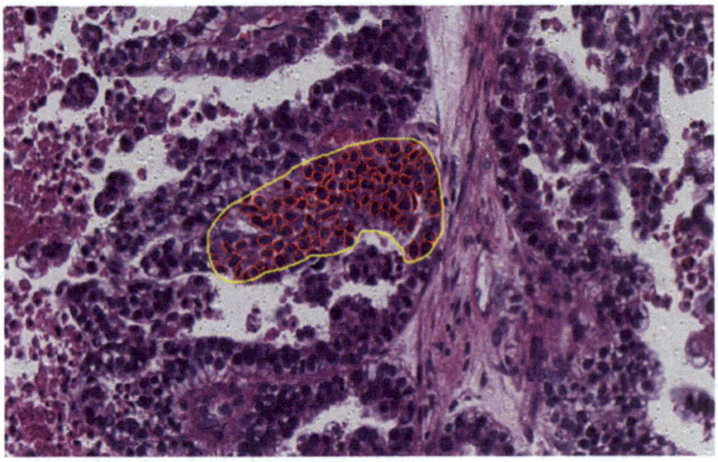

Fig. 1 Example of region of interest manually drawn with the QuPath software to include roughly 100 cells (based on nucleus segmentation). This image shows an H&E-stained slide of clear cell ovarian cancer, which was scanned with the Hamamatsu NanoZoomer high-resolution microscopy

3.4 Bioimage Analysis with QuPath

1. Import NDPI images generated with NanoZoomer into QuPath.
2. Identify ROIs (we chose patches of around 100 cancer cells from ovarian clear cell carcinoma; *see* Fig. 1) and three calibration points that will be used when importing the files generated with QuPath into the LCM (these structures need to be easy to identify at low magnification, i.e., 10× or 20×).
3. Apply InstanSeg extension to annotated ROIs to include a similar number of cells in each region: by applying the nucleus segmentation function, the extension will also provide the number of nuclei included in each ROI (*see* **Note 4**).
4. Export the GeoJson file containing the coordinates of the ROIs and the calibration points that have been identified with QuPath.

3.5 Removal of Coverslip and Drying the Slides

1. Soak the slides for at least 1 h in xylene (potentially overnight if the slides remained covered for more than a few hours; *see* **Note 5**).
2. Remove the coverslip.
3. Replace xylene by quickly dipping the slides in liquid xylene to remove any excess material remaining after coverslip removal, then dip twice in 100% ethanol for 2 min each, followed by two 1-min immersions in 80% ethanol, two 1-min immersions in 50% ethanol, a single 1-min immersion in 80% ethanol, and finally a 1-min immersion in 100% ethanol (*see* **Note 6**).

4. Leave the slide at room temperature for around 1 h to allow for evaporation of ethanol.
5. Store slides at 4 °C.

3.6 Optional: Draw Calibration Markings on the Slide

Perform the following steps if unable to select three easily identifiable calibration points when performing bioimage analysis with QuPath (*see* Subheading 3.4). This section will allow users to draw three calibration markings on the slide, which can then be used as calibration points in QuPath and will be easily recognizable when the slide is inserted into the LCM microscope for dissection of the ROIs (*see* **Note 7**).

1. Draw three calibration markings by inserting the slide in the LCM microscope. Use a 10× or 20× magnification for ease of recognition and use the following laser settings: 10× magnification, power 59, aperture 1, speed 40, middle pulse count 2, final pulse −1, head current 45%, pulse frequency 2980.
2. Perform a second scanning of the slide (this time with the calibration markings drawn on the slide and without the coverslip on).
3. Import the NDPI file corresponding to this image into QuPath.
4. Import the GeoJson file with the annotations generated on the image of the slide with the coverslip on (*see* Subheading 3.4).
5. Select calibration points that coincide with the markings drawn on the slide. This will allow easier recognition of the calibration points when performing LCM.
6. Select all the annotations with the following commands: Objects → Annotations → Transform Annotations. This will allow to drag all the annotations at once. Then adjust the position of the ROIs initially identified on the image with the coverslip on to precisely align with tissue structures on the new image (with the calibration markings drawn and without the coverslip).
7. Export the new Geojson file generated.

3.7 QuPath to LCM

1. Use the web app from Coscia's group to convert format from GeoJson to xml (https://qupath-to-lmd-mdcberlin.streamlit.app/).
2. Import the GeoJson file.
3. Follow the instructions on GitHub to assign each ROI to a specific well on the 384-well plate (*see* **Note 8**).
4. Run the app and download the contours and xml file containing the coordinates of calibration points and ROIs.

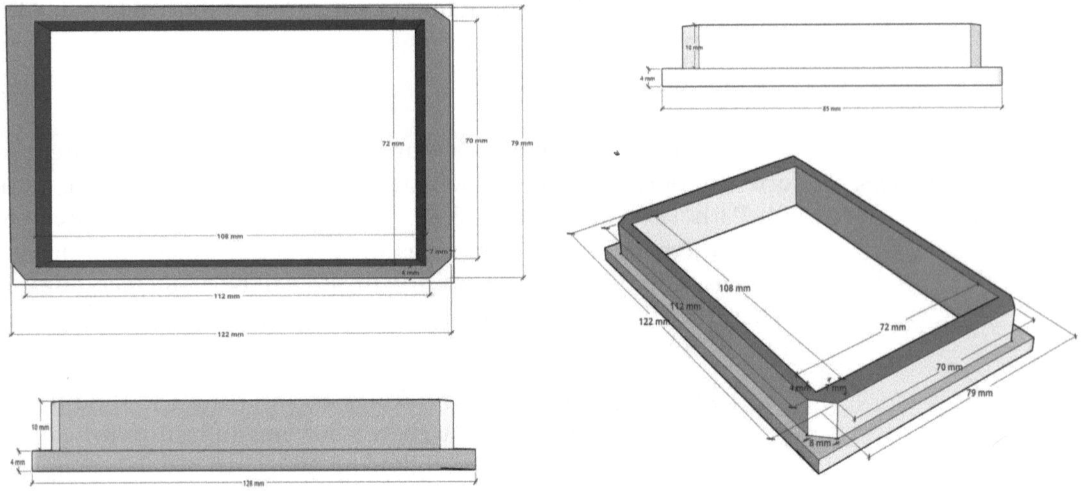

Fig. 2 Insert for LCM microscope LMD7 by Leica. This insert is placed under the collection plate and it allows to lift the plate 4 mm closer to the slide. The addition of an insert significantly reduces the dropout rates when performing LCM with LMD7. This figure was created using the SketchUp Modeling Software based on the insert produced by the Institute of Genetics and Cancer workshop

3.8 LCM

1. Optional: Use an antistatic gun on the 384-well plate to reduce electrostatic interactions (*see* **Note 8**).
2. Insert the adaptor (4.0 mm thick; *see* Fig. 2) into the plate holder and position the plate onto the adaptor.
3. Import the xml file into the Leica LCM system as per manufacturer's instructions.
4. Match the calibration points required by the LCM system with the calibration points selected during previous steps (*see* Subheading 3.4 or, optionally, Subheading 3.6).
5. Import the shapes.
6. Perform a specimen overview.
7. Change to an appropriate magnification based on the size of the ROIs to microdissect (we have used 40× magnification for maximal precision).
8. Perform cutting with the middle pulse option:
   (a) Settings: magnification 40×, power 59, aperture 1, speed 40, middle pulse count 2, final pulse −1, head current 45%, pulse frequency 2980 (these might need to be amended depending on the tissue and the size of the ROIs).
   (b) Up to two attempts at cutting per shapes were allowed.
   (c) Dissection was performed using the 10× 0.3 NA lens.
9. Inspect the wells to check for the presence of the sample.

Optional: Use a second adaptor of 7 mm thickness to elevate the plate further and allow the inspection of the entire well to ensure successful collection of the ROI (*see* **Note 9**).

10. Add 15 μL of ACN directly onto the wells containing the sample.
11. Centrifuge the samples for 1 min immediately. The centrifuge needs to be as close as possible to LCM to minimize the dropout rate.
12. Cover the 384-well plate with the cover lid.

    Optional: At this stage, it is possible to safely freeze the plate (*see* **Note 10**).

3.9 Digestion and Lysis

1. Dry the samples in a Speedvac for 20 min at 30 °C.
2. Add 2 μL of lysis buffer to each well and centrifuge for 1 min at 700 g.
3. Place in the thermocycler at 95 °C for 1 h.
4. Add 2 μL of digestion buffer to each well and centrifuge for 1 min at 700 g.
5. Place in the thermocycler at 37 °C for 8–10 h, then drop temperature to 4 °C if leaving overnight.

3.10 Liquid Chromatography Mass Spectrometry (LC/MS)

1. Stop the digestion via acidification of the solution with 20 μL of 1% TFA and centrifuge for 10 s.

   Optional: At this stage, it is possible to safely freeze the plate (*see* **Note 10**).
2. Load the samples from the 384-well plate onto Evotips according to the manufacturers' instructions.
3. To perform liquid chromatography/mass spectrometry, connect the Evosep 1 LC to a trapped ion mobility spectrometry quadruple time-of-flight spectrometer (such as the timsTOF SCP from Bruker; *see* **Note 11**).
4. Use a 40SPD Whisper Zoom (as designed by Evosep) method to separate the peptides.
5. For diaPASEF analysis, use a diaPASEF method with 12 diaPASEF scans separated into 2 ion mobility windows per scan covering a 300–1200 m/z and an ion mobility range from 0.6 to 1.6 Vs cm^{-2}. Optimal m/z and ion mobility windows were generated by py_DIAid (*see* Table 1) [11]. Operate the mass spectrometer in high-sensitivity mode, with an accumulation and ramp time at 100 ms, capillary voltage set to 1400 V, and the collision energy as a linear ramp from 20 eV at $1/K_0 = 0.6$ Vs cm^{-2} to 59 eV at $1/K_0 = 1.6$ Vs cm^{-2}. The collision energy was ramped linearly as a function of ion mobility from 59 eV at $1/K_0 = 1.6$ Vs cm^{-2} to 20 eV at $1/K_0 = 0.6$ Vs cm^{-2}.

Table 1
Optimal m/z and ion mobility windows generated by py_DIAid

Window #	m/z start	m/z end	Ion mobility start ($1/K_0$)	Ion mobility end ($1/K_0$)
1	400	425	0.636	1.204
2	425	450	0.659	1.227
3	450	475	0.683	1.251
4	475	500	0.707	1.275
5	500	525	0.73	1.298
6	525	550	0.754	1.322
7	550	575	0.778	1.346
8	575	600	0.801	1.369

3.11 Data Analysis

There are multiple options to perform data analysis on proteomic data. Mann's group has developed a Python-based platform known as AlphaPepStats [12], while other groups have used the Perseus software [13]. We use R packages to perform the following steps (the code has been made available on GitHub; *see* **Note 12**) [14]:

1. Data import.
2. Data cleaning to remove nonnumeric data.
3. Normalization for PCA: log-2 transformation.
4. Checking for normal distribution of data with Q-Q plot.
5. K-nearest neighbors imputation.
6. Principal component analysis.
7. K-means clustering to identify the optimal number of clusters.
8. Z-score of rows and columns.
9. Creation of a heatmap with hierarchical clustering.
10. Differential protein expression analysis with Limma package or ANOVA based on the number of clusters identified.
11. Generation of a volcano plot or box plot to illustrate differentially expressed proteins.

4 Notes

1. Bioimage analysis. The two software devices most commonly used by the spatial proteomics community are BIAS (Single-Cell Technologies) and QuPath [4, 10].

   QuPath is an open-source software that allows users to easily annotate ROIs with no coding requirements. It also allows nucleus and cell segmentation and has some features to apply machine learning-based models to perform pixel or object classification. Similarly, BIAS is a software aimed at bio-image analysis at single-cell level, which makes use of the most up-to-date AI methods.

When dealing with images of H&E-stained slides, QuPath has proven to be easier to use and effective in identifying ROIs containing around 100 cells. However, if the user is seeking to perform segmentation of individual cells (which can then be pulled into different wells according to their immunophenotype), BIAS currently outperforms QuPath in terms of in-built machine learning tools, and the possibility to create a well-structured pipeline for the whole set of slides.

2. Use of xylene-based mounting media to maximize image clarity. There are different options when it comes to the choice of appropriate mounting media for spatial proteomics work. Most groups opted for an aqueous, glycerol-based, mounting medium [4, 5]. The main advantage of this is the ease of removal of the coverslip after imaging, which can be done in just a few minutes by immerging the slide in PBS and leaving it on a shaker. In addition to this, some evidence also suggests that glycerol might act as a stabilizer of the PEN membrane and improve the ability to perform antigen retrieval for multiplex imaging [15]. However, not every glycerol-based mounting medium is adequate for downstream cutting with the LCM, and it is important to check the other components of the medium. For instance, some of these media are stored in solid form (in the fridge) and need to be liquefied before use. Removing the coverslip after use of such media comes with a higher chance of damage to the PEN membrane, and the use of higher temperatures to liquefy the solution can remove eosin from the stained slides.

   To compensate for the lack of surface markers that allow to clearly distinguish different cell types and structures within a tissue, when using H&E-stained samples, it is important to maximize the clarity of the image obtained with high-resolution microscopy. To do this, we decided to use a xylene-based mounting medium, instead of an aqueous one, as this is the gold standard for H&E-stained slides. As shown in Fig. 3, the use of this type of media has significantly increased the definition of the images obtained with the NanoZoomer when compared to those generated when using a glycerol-based medium.

3. Whole-slide imaging systems. Several systems are currently available for whole-slide imaging, including NanoZoomer, Axioscan, and LAS X (which is the integrated software of the LMD Leica microdissection microscope). In our experience,

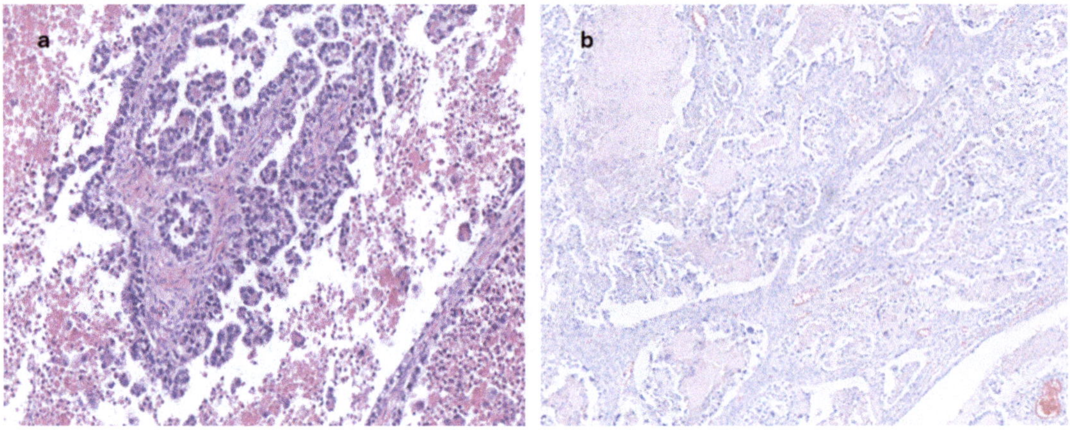

Fig. 3 Comparison of image clarity of H&E-stained slides of clear cell carcinoma of the ovary with different mounting media. A xylene-based mounting media was used on the right (**a**) whereas a glycerol-based approach was used on the right (**b**)

the NanoZoomer has proved to be quicker (around 4 min to scan a slide compared to 8–10 min with the other systems) and more intuitive (due to ease of definition of the scanning region and freedom of choice of focus points) when compared to both LAS X and the Axioscan.

4. Nucleus segmentation in QuPath. Another important consideration when working with ROIs comprised of patches of cells (and not individually dissected cells) is the need to have a similar amount of material in each well after collection. To do this, the InstanSeg extension [16] in QuPath is a deep learning-based method that allows to perform fast and reliable nucleus segmentation of H&E images (and potentially IF) and can calculate the number of cells per region, as illustrated in Fig. 4. This allows users to consistently select ROIs comprised of a prespecified number of cells, which is essential when performing MS-based proteomics.
5. The main challenge of using xylene-based mounting media for spatial proteomics is the removal of the coverslip. In fact, xylene-based mounting media need to solidify before safely introducing the slide into a whole-slide imaging system (by leaving the slides for around 1 h at room temperature). Following this, removal of the coverslip can be obtained by soaking the slide in xylene. This usually takes around 1 h if the slide has been covered and imaged quickly (i.e., within a few hours). However, if the slide has remained covered for many hours, or days, it is then necessary to soak the slide in xylene overnight. After removal of the coverslip, it is necessary to remove the residual xylene prior to LCM. This step is crucial as failure to do so would prevent the laser from dissecting the ROIs from the tissue and collecting these into a well plate (as detailed in **Note 6**).

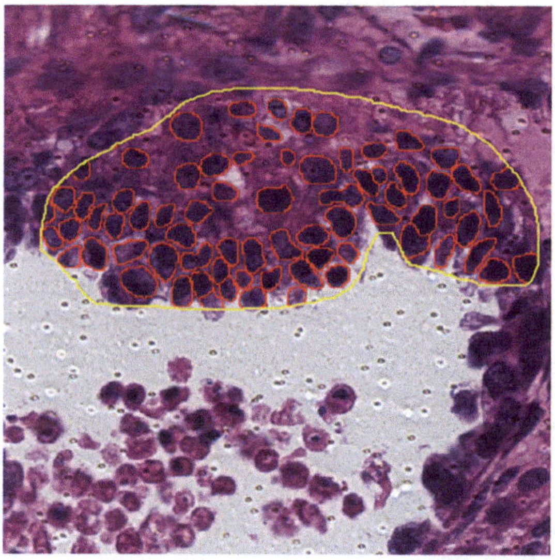

Key	Value
Image	22236_28714.ndpi
Object ID	95046b83-99ce-4cba-8cc8-515a7...
Object type	Annotation
Name	
Classification	
Parent	Root object (Image)
ROI	Polygon
Centroid X µm	16232.5521
Centroid Y µm	4095.4514
Num Detections	74
Area µm^2	7199.2473
Perimeter µm	398.4171

Fig. 4 The InstanSeg extension in QuPath uses deep learning to perform nuclear segmentation on H&E-stained slides. This extension was used to estimate the number of cells included in each ROI, hence allowing to select regions of similar sizes and with comparable numbers of cells. The ROI shown on the left side of the figure contains 74 detections, and the table on the right was automatically generated by InstanSeg to illustrate the number of cells included, the area of the ROI, and its coordinates

6. Comparison of drying process of slides mounted with glycerol-based and xylene-based media. If the slide has been mounted with a glycerol-based mounting medium, after removal of the coverslip by gentle agitation in PBS for 5 min on a shaker, the slide can be properly dried by dipping it into increasing concentrations of ethanol (10 dips in 50%, followed by 10 dips in 80%, 10 dips in 100%, and 10 dips in 100% ethanol again) [5]. After this, the slides can be left to complete the drying process at room temperature for around 1 h before performing LCM. However, as xylene is not miscible with water and only partially miscible with ethanol, the process is slightly different if using a xylene-based mounting media. After removal of the coverslip by soaking in xylene, the slides are then quickly dipped in xylene to remove any remaining solid component, and then are left for 2 min in 100%, 80%, 50% ethanol, respectively. At this point, xylene has been replaced by a mixture of ethanol and water, which is then removed by dipping the slides for around 1 min in 80% ethanol and 100% ethanol, and by completing the drying process at room temperature for around 1 h.
7. Calibration markings. When importing ROIs annotated with QuPath into the LCM, the Leica software requires users to identify three calibration points as a reference to calculate the coordinates of the ROIs. The user has to then recognize the

calibration points used in QuPath on the slide when loaded into the laser microdissector (n.b., the slide is now not covered by a coverslip and image clarity is far inferior to what obtained with whole-slide imaging systems like the NanoZoomer). Calibration points can thus be quite challenging to recognize when visualizing the slide on the LCM, and it is important for users to select points easily distinguishable even at lower magnifications. Depending on the heterogeneity of the tissue being investigated, it might be necessary to draw calibration markings with the LCM itself after removal of the coverslip (and appropriate drying of the slide), which will then be easier to spot when the slide is reinserted into the LCM. The timing is crucial: If the calibration markings are drawn before applying the coverslip or with the coverslip already on the slide, there is a significant chance of liquid (i.e., the mounting medium) penetrating through the markings underneath the PEN membrane, which will then prevent the laser from dissecting the desired ROIs (even if removal is performed as detailed above). Thus, it is pivotal that the markings are drawn after removal of the coverslip. After adding the calibration markings, the slides need to be re-scanned (without the coverslip on), and the annotations previously created in QuPath imported into another QuPath file (within the same project) with the uncovered image. This will allow users to select as calibration points the markings drawn on the slide with the LCM, which will significantly aid with identifying these points when the slide is inserted back into the LCM for microdissection. It is possible that the ROIs identified on the image with the coverslip on will be slightly misaligned with tissue structures visualized on the uncovered slide (as the scanning areas automatically identified by the NanoZoomer might differ on the two images, usually by only a few millimeters). Users will then need to adjust the positioning of all the ROIs created, which can be done en bloc for all the annotations in QuPath with the following functions: Objects → Annotations → Transform annotations.

8. From QuPath to LCM. Coscia's group has created a Streamlit Web App to convert the GeoJson file created with QuPath to a Leica LMD-compatible xml file, which also includes the wells of a 384-well plate where each ROI will be dissected into. It is important to note that sometimes QuPath labels irregularly shaped regions as "multipolygons" (instead of "polygon"). The web App is currently unable to deal with "multipolygons" and compatibility issues will arise when trying to import the xml file. For this reason, before converting to xml with the web app, it is important to inspect each ROI created to make sure it is a "polygon" and not a "multipolygon."

9. Collection in well plate after LCM microdissection. An important aspect to consider when using LCM is that there are multiple factors that affects falling of the ROIs into the expected wells. These include airflow, electrostatic interactions, and humidity. For this reason, it is advisable to strictly monitor room temperature and humidity (45% or less), and it is conceptually helpful to also use an antistatic gun right before cutting. In addition to this, on the Leica LMD7, we found it useful to add an adaptor inside the plate holder, which lifts the plate closer (by 4 mm) to the slide (*see* Fig. 2; the plate shown is only compatible with Eppendorf LoBind collection plate). In addition to this, a second adaptor could be used to move the plate 7 mm closer to the slide, which will allow inspection of the entirety of the well with the 40× lens to ensure a successful collection of the tissue in the desired well.
10. Freezing the samples. Since the framework here presented includes multiple, labor-intensive steps, it is important to consider whether users could safely stop at certain points, freeze the samples, and continue with the experiment later. It is indeed possible to store the well plate in −80 °C either after collection of the samples into the wells (i.e., after Subheading 3.8, **step 12**) or after stopping protein digestion with TFA (i.e., after Subheading 3.10, **step 1**).
11. Mass spectrometry-based proteomics expected throughput. Our group uses a timsTOF SCP instrument (Bruker) and the py_diAID (Automated Isolation Design for Data-Independent Acquisition) Python package to generate data-independent acquisition methods. We have used the dia-PASEF scan mode [11]. On clinical samples derived from patients with clear cell ovarian carcinoma, we consistently identified around 5000 proteins from regions comprising around 100 cells, and we were able to process 40 samples per day. Different throughput will be achieved depending on the tissues type, the MS instrument, and the scanning parameters used.
12. Bioinformatics analysis. There are several options that have been used by the spatial proteomics community, including Perseus [13], AlphaPepStats [12], and R packages.

    AlphaPepStats is an open-source Python package that can analyze data generated with AlphaPept, DIA-NN, FragPipe, and MaxQuant [12], and its main advantage is to provide a structured workflow to import, preprocess, and visualize data.

    Similarly, Perseus is an R-based platform that contains statistical tools to analyze high-dimensional omics data, and its main advantage is the user-friendly interactive environment that allows intuitive usability [13].

Finally, bioinformatics analysis might be conducted with R. The main advantages of this platform are the availability of a large community of developers and users, its flexibility (it allows to customize the analysis with bespoke scripts and thousands of packages), and the widespread availability of documentation on how to use statistical tools on this platform. Regardless of the chosen method, it is pivotal that the pipeline used contains adequate normalization and imputation processes, and statistical tests appropriately selected based on the distribution of the data. This can be done with the in-built resources available with Perseus and AlphaPepStat or with the code made available on GitHub from our group [14].

Acknowledgments

This method was developed in collaboration with multiple members of the Institute of Genetics and Cancer, University of Edinburgh. In particular, I am grateful to Prof. Gourley for his guidance, to the Advanced Imaging Resource (Dr. Ann Wheeler, Matthew Pearson, and Martin Lee) for their assistance with optimizing LCM, to the Institute of Genetics and Cancer workshop for creating the insert for the LCM microscope, to the histology department (Ian Croy and Helen Caldwell) for their expertise in tissue mounting, staining, and high-resolution microscopy, and to Dr. John Thomson for his support with bioinformatics analysis.

References

1. Guo W, Hu Y, Qian J et al (2023) Laser capture microdissection for biomedical research: towards high-throughput, multi-omics, and single-cell resolution. J Genet Genomics 50: 641–651
2. Mann M, Kumar C, Zeng W-F et al (2021) Artificial intelligence for proteomics and biomarker discovery. Cell Syst 12:759–770
3. Gordon A, Gousset K (2021) Utilization of laser capture microdissection coupled to mass spectrometry to uncover the proteome of cellular protrusions. In: Carrera M, Mateos J (eds) Shotgun proteomics. Springer US, New York, pp 25–45
4. Mund A, Coscia F, Kriston A et al (2022) Deep visual proteomics defines single-cell identity and heterogeneity. Nat Biotechnol 40:1231–1240
5. Makhmut A, Qin D, Fritzsche S et al (2023) A framework for ultra-low-input spatial tissue proteomics. Cell Syst 14:1002–1014.e5
6. Arreaza G, Qiu P, Pang L et al (2016) Pre-analytical considerations for successful next-generation sequencing (NGS): challenges and opportunities for formalin-fixed and paraffin-embedded tumor tissue (FFPE) samples. Int J Mol Sci 17:1579
7. Strom SP (2019) Fundamentals of RNA analysis on biobanked specimens. In: Yong WH (ed) Biobanking. Springer New York, New York, pp 345–357
8. Coscia F, Doll S, Bech JM et al (2020) A streamlined mass spectrometry–based proteomics workflow for large-scale FFPE tissue analysis. J Pathol 251:100–112
9. Wu W, Choudhry H (2013) Next generation sequencing in cancer research: volume 1: decoding the cancer genome. Springer New York, New York
10. Bankhead P, Loughrey MB, Fernández JA et al (2017) QuPath: open source software for digital pathology image analysis. Sci Rep 7:16878

11. Meier F, Brunner AD, Frank M et al (2020) diaPASEF: parallel accumulation–serial fragmentation combined with data-independent acquisition. Nat Methods 17:1229–1236
12. Krismer E, Bludau I, Strauss MT et al (2023) AlphaPeptStats: an open-source Python package for automated and scalable statistical analysis of mass spectrometry-based proteomics. Bioinformatics 39:btad461
13. Tyanova S, Temu T, Sinitcyn P et al (2016) The Perseus computational platform for comprehensive analysis of (prote)omics data. Nat Methods 13:731–740
14. Gasparello T (2014) SpatialProt, GitHub repository. https://github.com/vonkriegsheim/SpatialProt/tree/main
15. Nordmann TM, Schweizer L, Metousis A et al (2023) A standardized and reproducible workflow for membrane glass slides in routine histology and spatial proteomics. Mol Cell Proteomics 22:100643
16. Goldsborough T, Philps B, O'Callaghan A et al (2024) InstanSeg: an embedding-based instance segmentation algorithm optimized for accurate, efficient and portable cell segmentation. arXiv:2408.15954. https://arxiv.org/abs/2408.15954

Chapter 15

Automated Intelligent Microscopy for Phenotype Identification, Spatial Localization, and Retargeting of Cells

Simone Pelicci, Laura Furia, Francesco Spadari, Roberto Grigolato, Luca Calabrese, Ilaria Girolami, and Mario Faretta

Abstract

Heterogeneity is a hallmark of cancer progression, characterized by genetic, transcriptional, and protein expression variability, as well as dynamic interactions within the tumor microenvironment. Primary tumors often harbor multiple, spatially distinct cellular subpopulations, complicating efforts to achieve comprehensive molecular characterization. Recent progress in molecular diagnostics and sequencing technologies has played a pivotal role in revealing the complex genetic architecture of various cancers. While Next-Generation Sequencing (NGS) has enhanced molecular profiling, traditional bulk approaches obscure spatial and single-cell resolution. Emerging spatially resolved and single-cell technologies allow the investigation of tumors within their tissue architecture. Among these, Laser Capture Microdissection (LCM) enables precise isolation of defined regions for downstream multi-omics analysis, preserving spatial context. This protocol offers an automated analysis-driven microscopy workflow for spatial-omics profiling. By leveraging high-content imaging, with scalable resolution, and A.M.I.CO. image cytometry platform, it enables phenotype-driven image acquisition and precise spatial targeting across diverse sample types. The spatial relocalization procedure integrated in A.M.I.CO. provides the physical coordinates of phenotypically defined cells or regions inside a sample for LCM. This protocol supports an intelligent analysis-driven acquisition, offering a reproducible, scalable, and tissue-context-aware workflow for several biomedical applications.

Key words High-content microscopy, Scalable resolution, Spatial relocalization, Image cytometry, Spatial biology

1 Introduction

Cancer is an inherently heterogeneous disease, marked by considerable variability in its genetic, transcriptional, and protein expression landscapes. This heterogeneity is not only related to a clonal evolution of genetically distinct populations of tumor cells but also to the intricate and dynamic interplay between tumor cells and the surrounding microenvironment [1–3]. As such, understanding cancer biology today requires the investigation of both the tumor's

Roberta Noberini (ed.), *Laser Capture Microdissection: Methods and Protocols*, Methods in Molecular Biology, vol. 3015, https://doi.org/10.1007/978-1-0716-5154-4_15,

molecular signature and its spatial context within the tissue architecture [4]. Technological advancements, such as next-generation sequencing (NGS), and single-molecule proteomics have transformed our ability to profile tumors at an unprecedented molecular depth [5, 6]. However, traditional bulk sequencing techniques average signals across heterogeneous cell populations, thereby hiding critical insights into individual cell states and their spatial relationships [7, 8]. This averaging effect limits our ability to decipher tumor heterogeneity, track clonal evolution, and understand the localized influence of the microenvironment. To overcome this obstacle the research community has increasingly turned to single-cell and spatially resolved technologies, which provide both molecular detail and spatial localization [9–11].

Despite the evolution of single-cell technologies, studies of target cell populations of low-medium frequency rely on sample enrichment. Within this landscape, Laser Capture Microdissection (LCM) represents a cornerstone method for linking morphological context with molecular analysis. Unlike bulk tissue processing, which homogenizes signals, LCM facilitates downstream molecular profiling (e.g., DNA/RNA sequencing, proteomics, methylation analysis) from spatially defined regions, making it a powerful tool for studying intratumoral heterogeneity and microenvironmental influences [12, 13]. Moreover, LCM is highly compatible with Formalin-Fixed Paraffin-Embedded (FFPE) samples and cryosections, enabling retrospective analyses of archival tissues [14, 15]. Its integration with downstream single-cell and spatial transcriptomic workflows enhances our ability to correlate histological features with molecular phenotypes [16].

Despite these innovations, challenges remain linked to the high degree of subjectivity related to the target identification, typical of man-driven techniques. To remove these limitations, automation has been introduced at various stages of the acquisition and analysis pipelines. Advances in automated microscopy and image analysis have revolutionized the way biological samples are studied, offering unprecedented scalability, reproducibility, and quantitative insights [17]. Commercially available motorized fluorescence microscopes can acquire high-resolution images across large tissue sections with minimal user intervention, while software, often open-source, enables reproducible image registration, segmentation, and data quantification [18, 19]. The incorporation of analysis-driven acquisition strategies allows researchers to modulate resolution and imaging parameters based on biological relevance, thereby optimizing data storage and time consumption [20, 21]. Modern workflows now integrate high-throughput imaging platforms with sophisticated computational pipelines capable of handling diverse sample types, including cultured cells, histological tissue sections, organoids, and Tissue Micro-Arrays (TMAs) [22]. The integration of automated microscopy with image cytometry [23] and machine learning has transformed the interpretation of image data. Platforms such as CellProfiler [24], Ilastik [25], and QuPath [26], along with deep neural network-based models, have facilitated the transition from qualitative visual inspection to quantitative, data-driven discovery. This paradigm shift enhances reproducibility and objectivity, opening avenues for large-scale image-based profiling, predictive modeling, and digital pathology. However, despite significant technological progress, implementing a robust and versatile microscopy pipeline that can adapt to varying sample morphologies and staining protocols remains a complex task requiring domain-specific expertise.

In this chapter, we will provide a detailed description of an optimized protocol to perform an automated in situ spatial-omics analysis and efficiently relocate areas of interest, corresponding to specific cellular phenotypes, to be selected for LCM. We focus on critical steps including image acquisition, preprocessing, segmentation, and analysis. Particular emphasis is placed on the challenges and solutions associated with handling heterogeneous biological materials: from monolayer cell cultures to complex 3D structures such as tissue sections and TMAs. This protocol outlines the analysis-driven acquisition strategy routinely used in our lab [22, 27, 28] that integrates flexible, re-scalable image resolution with our image cytometry analysis software (Automated Microscopy for Image CytOmetry (A.M.I.CO.) [29]). By linking acquisition directly to the experimental question, the protocol enables researchers to adapt acquisition parameters scaling from widefield imaging of large areas to high-resolution and sensitivity for subcellular analysis, thanks to the employment of high Numerical Aperture (NA) oil immersion objectives. Image cytometry analysis identifies and characterizes phenotypes of interest across complex biological samples, thus allowing an intelligent selection of targets, based on quantitative, image-derived data, rather than a random or human-driven selection. A core component of this protocol is the spatial relocalization process. Utilizing an integrated cell retrieval procedure based on the hardware metadata, the spatial *xy*-coordinates of target cells or specific regions are stored and subsequently converted into microscope stage positions, identified in earlier analysis steps. This functionality is essential for two primary reasons: (1) It enables the identification and localization of regions of interest (ROIs) within the sample, by leveraging low-resolution (dry objective) images of large areas, previously collected in the earlier steps of the acquisition pipeline, and (2) it provides single-cell relocalization of phenotypically defined cells for targeted reacquisition with different imaging modalities (e.g., super resolution microscopy) and/or laser capture microdissection (LCM), thereby supporting spatial-omics profiling of selected cellular populations or tissue microenvironments. This re-scalable resolution approach, combined with the cell relocalization process, optimizes acquisition time, storage, and statistical sampling, preserving the ability to capture detailed, high-content data and minimizing unnecessary data collection. This protocol aims to provide a practical guide for researchers seeking to implement or refine automated imaging systems across various research applications.

2 Materials

2.1 Sample Preparation Reagents

Sample preparation can be performed following a standard immunostaining protocol.

2.2 Microscope Setup

The following protocol can be employed with any motorized fluorescence microscope, making minimal modifications. Here, we list the equipment:

1. Inverted Nikon Eclipse Ti2 microscope (Nikon instruments, Tokyo, Japan) equipped with an A1R confocal scanhead, Structured Illumination (SIM), and N-STORM modules (Nikon instruments, Tokyo, Japan). The fully motorized automated microscope is controlled by the NIS Elements software (version 5.42.07).
2. Excitation sources:
   (a) LED light source (pE-4000, CoolLED, Andover, UK) with 16 selectable wavelengths for widefield microscopy.
   (b) Laser unit (LU-NV, Nikon instruments, Tokyo, Japan) equipped with five laser lines (405 nm (23.1 mW), 440 nm (25.5 mW) 488 nm (79.1 mW), 561 nm (79 mW), 647 nm (137 mW)) for confocal microscopy.
   (c) Laser bench (L4Cc combiner, Oxxius S.A., Lannion, France) equipped with four high-power sources (405 nm (216 mW), 488 nm (240 mW), 561 nm (240 mW), 640 nm (360 mW)) and two acousto-optic modulators for single-molecule microscopy.
3. Optical filter sets: A multiband dichroic mirror (C-NSTORM QUAD 405/488/561/647 FILTER SET; Chroma) and an emission filter wheel (OptoSpin, Cairn Research Ltd., Faversham, Kent, UK) are equipped to filter the light. The physical focus stabilization is guaranteed by the reflection of an infrared diode module (Perfect Focus System, Nikon instruments, Tokyo, Japan).
4. High-precision XY motorized microscope stage for automated acquisition of multiple regions of interest (ROIs) around the sample.

5. CMOS camera (Dual ORCA Flash 4.0 Digital CMOS camera C13440, Hamamatsu, Tokyo, Japan) to detect the fluorescence signal.
6. Apochromatic lenses for optimized correction of the optical aberrations.
7. List of objectives according to the sample features and the experimental step:
   (a) Low magnification objectives (4X, 10X, 20X): (i) Fast scanning of large area of the sample (millimeters), (ii) limited data storage, and (iii) single-cell detection at low resolution.
   (b) High magnification objectives (60X, 100X): (i) High-resolution and/or high-sensitivity collection of images, (ii) detection of intracellular details at single-cell level, (iii) single-molecule analysis when required (super resolution microscopy).

2.3 Software

1. NIS Elements software version 5.42.07 (Nikon instruments, Tokyo, Japan).
2. Automated Microscopy for Image CytOmetry (A.M.I.CO.) analysis package (https://github.com/MarioFaretta/AMICO) developed in the ImageJ macro language (ImageJ, version 1.54j, Rasband, W.S., US National Institutes of Health, Bethesda, Maryland, USA, https://imagej.net/ij/).

3 Methods

3.1 Sample Preparation and Staining for Imaging

Given the large number of experimental models in cancer research, the pipeline is characterized by a high degree of flexibility. The workflow can be adapted to fixed 2D cell culture, frozen and FFPE tissue sections, spheroids, in vitro patient-derived organoids, and biopsies from animal models including patient-derived xenografts.

For sample preparation, the user can adopt any established staining protocols for immune-fluorescence microscopy.

Generally, samples are stained with DAPI (or Hoechst 33342) to visualize nuclei and secondary antibodies conjugated with different fluorophores to visualize all the antigens of interest (e.g., Pacific Blue, Pacific Orange, Alexa Fluor 488, Cy3, Alexa Fluor 555, and Alexa Fluor 647). For cell cycle analysis on cultured cells, Click-iT EdU Pacific Blue assay kit (C10418, Thermo Fisher Scientific, Waltham, MA, USA) is normally employed to measure DNA synthesis (*see* **Note 1**).

Important Choosing the correct spectral range for imaging in tissue samples is essential for minimizing autofluorescence and optimizing detection efficiency. Autofluorescence can interfere with the signal from specific fluorophores, making the acquisition process more challenging. This unwanted background signal generally occurs in the visible spectral range (typically between 400 and 600 nm). Therefore, a preliminary screening of the unstained samples is strongly suggested to optimize fluorophore selection, excluding as much as possible the highest autofluorescence range. One effective strategy to reduce autofluorescence is to utilize fluorophores that emit in the near-infrared (NIR) range, generally between 700 and 900 nm. At these wavelengths, autofluorescence is almost absent, resulting in a cleaner signal and better contrast. However, using the NIR range presents certain challenges. NIR fluorophores tend to have lower emission efficiency compared to those in the visible spectrum. Additionally, sensors and cameras have a dramatic drop in their light collection efficiency in this spectral range.

3.2 Image Acquisition

3.2.1 Preliminary Imaging Settings

1. Turn on the imaging setup.
2. Position the sample on the microscope stage with the proper support.
3. Choose the optimal objective based on the type of sample (e.g., cells, tissue, or organoids) and the experimental target (cell population selection or high-resolution and high-sensitivity analysis (e.g., RNA spots detection)) (*see* Subheading 3.2.2, **step 5** for further details) (*see* **Note 2**).
4. For every fluorescence target:
   (a) Define the optical configuration: Excitation source (LED or laser), excitation, and emission filter bandwidths.
   (b) Optimize image acquisition parameters: Set excitation source power and optimal exposure time to maximize dynamic range (e.g., greater than 1000 grayscale units for 16 bit-depth image, corresponding to 65,535 (2^{16}) intensity classes) and avoid saturation and photobleaching.

3.2.2 Automated Acquisition Workflow

Data acquisition is managed by the microscope control software. In our system, the acquisition protocol was coded into the JOBs plugin module of the microscope control software. Here, we outline the general procedure to design an automated imaging experiment (Fig. 1). The following steps are coded in our automated imaging workflow:

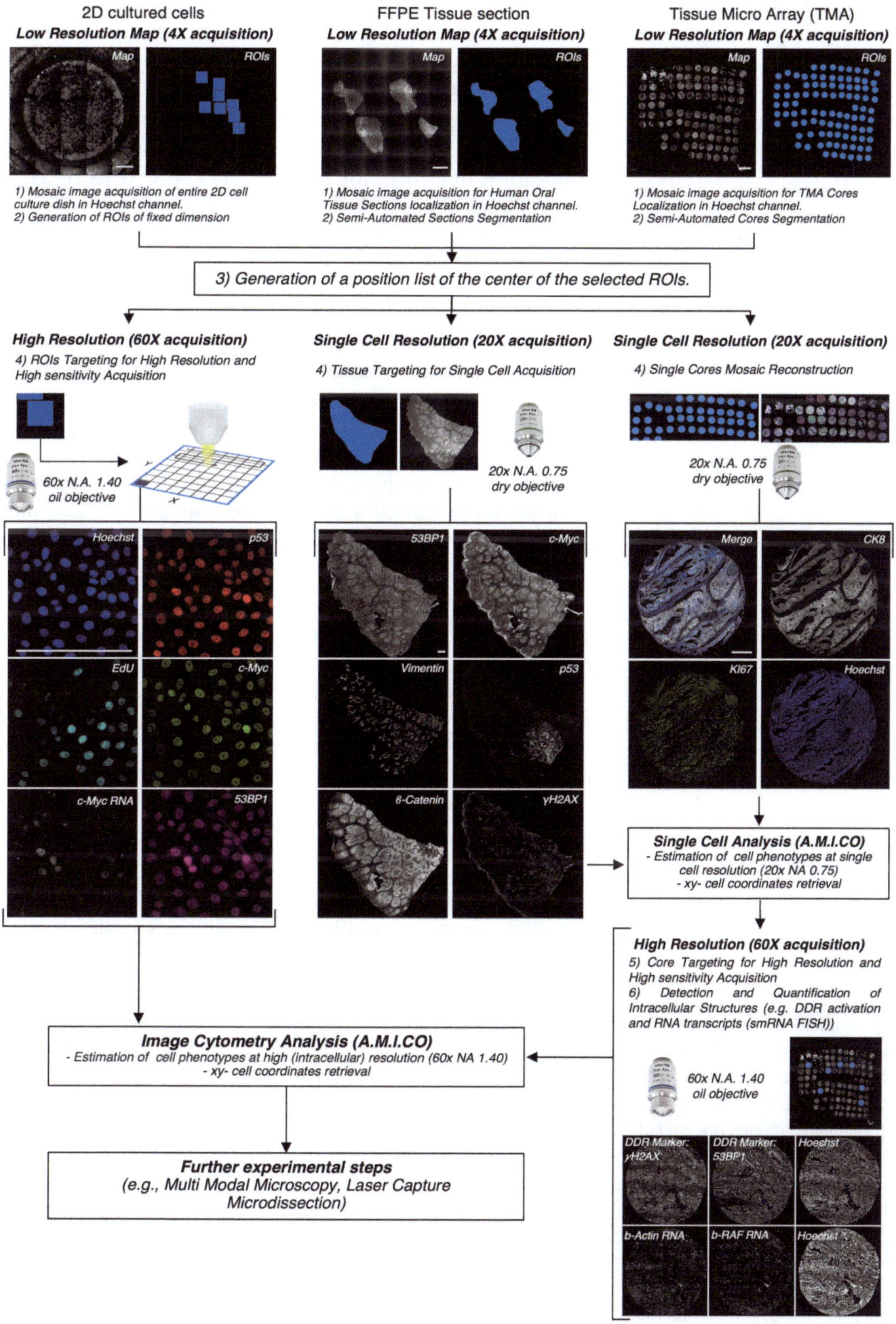

Fig. 1 Semiautomated Multi-Resolution Workflow for the spatial target of specific phenotypes. **Steps 1–2**. Samples (2D cultured cells, FFPE tissue section, Tissue Micro Array (TMA)) are acquired by a 4×/N.A. 0.13, and the resulting image (Map) is used to locate the regions of interest (ROIs). Scale bar: 2 mm. **Step 3**.

1. Check image acquisition parameters, previously defined in Subheading 3.2.1 before launching the acquisition routine of the control software (manual).
2. Acquire an overview of the sample (slide or dish) by a low magnification objective, for example, 4×/N.A. 0.13 or 10×/N.A. 0.5, according to the resolution required by the target selection, usually employing the DNA-dye channel. Image stitching will be then executed to reconstruct the entire field of interest.

Important The acquisition workflow starts with the fast construction of a low magnification map to target the regions of interest (ROIs) for the successive acquisitions at increased resolution. When acquiring cell cultures, target areas can be defined to avoid regions containing clustered cells, areas with optical defects, or with features that can compromise autofocus research.

With histological samples, location on the slide of multiple tissue sections can be stored for successive acquisition. According to the same procedure, cores of a Tissue Micro-Array (TMA) can be localized.

3. Define the ROIs from the low magnification map (manual or semiautomated).

   The ROIs can coincide with (i) the entire 2D cell culture sample or tissue slice(s) or can be confined to a specific histological area in the sample (ii) multiple areas of 2D or 3D sample.

   Practical examples:

   (a) *Histological tissue section*: The mosaic image of entire tissue slice is used to define a ROI for the successive acquisitions at increased resolution. The ROI can coincide with a specific morphological area in the sample.

Fig. 1 (continued) The corresponding masks with the positions of the ROIs are then stored and opened by the microscope control software to set the stage coordinates for the next image collection. **Step 4**. The selected ROIs of the samples are relocated and images acquired as the result of a multipoint acquisition. Every ROI is acquired by collecting adjacent positions, covering the entire field of interest. Every channel is acquired sequentially. Images are collected at single-cell resolution with a 20× 0.75 NA dry objective and/or at higher resolution with 60× N.A. 1.40 oil immersion objective, for the detection of the intracellular details. As shown in the TMA acquisition workflow, the 20X acquisition is eventually followed by an intermediate analysis-driven step by A.M.I.CO. software that allows the selection and identification of target phenotypes at single-cell-resolution data (e.g., single cores targeting). Image coordinates of the cells are then reconverted into microscope stage positions for the next acquisition at high resolution. Scale bar: 200μm. **Steps 5–6**. Finally, A.M.I.CO. is employed to perform image cytometry analysis to calculate single-cells statistics according to a classical flow cytometry interface. The software performs additional analysis (e.g., subcellular analysis) and spatially relocates targeted cells on microscope stage for further experimental steps (e.g., laser capture microdissection)

(b) *Tissue Micro-Array (TMA)*: The mosaic image of the entire TMA is used for a preliminary observation of the cores grid. Single cores were segmented and the corresponding mask with the cores position was then stored for the next image collection.

(c) *2D cell culture*: The mosaic image of the entire dish (or plate) is used to locate ROIs of fixed dimension to be acquired. This way, regions with optimal cell density and free of defects (e.g., dust, antibody precipitates, and large cell aggregates) could be excluded from the acquisition.

4. Generate a position list centered on the selected ROIs.
5. Choose acquisition modality (manual):

(a) *Cell resolution imaging (20× N.A. 0.75 dry objective).*

The images collected at low magnification provide whole-cell resolution data. The numerical aperture allows the quantification of the total amount of expressed markers (e.g., cell marker positivity in 2D cell culture and 3D samples (tissue sections, spheroids, organoids)) for a general identification of cellular subpopulations. Low magnification objectives provide a statistical and spatial overview of protein profile of highly expressed markers in the entire sample with an optimal data sampling and duration of the imaging. They are frequently employed as an intermediate step in acquisition of cell-resolved images to target a complex phenotype in histological samples, tissue sections, or TMA.

(b) *High-resolution imaging (60× N.A. 1.4 oil immersion objective).*

High-resolution widefield acquisition of images is almost mandatory when profiling the number and intensity of intracellular spatial details (e.g., foci, spots, nucleoli) or to perform high-sensitivity analysis of biomolecule content. It is worthwhile to remind that increasing NA (more than 1), by using immersion objectives, not only provides higher spatial resolution but also dramatically enhances signal sensitivity. Thus, analysis of the results can be greatly facilitated by the high signal-to-noise ratio when low expressed antigens are targeted. More than this, performing experiments of in situ transcriptomics to detect single RNA molecules (smRNA) by FISH requires high sensitivity and high spatial resolution to precisely measure the number of targeted transcripts and reduce errors in the correct detection, localization, and assignment of the RNA spots (*see* **Note 3**).

6. For every region, a two-step autofocus procedure is employed based on the DNA-dye signal. The microscope stage first moves to the center of each ROI and performs a wide-range search (from 10 to 30μm according to the sample thickness and linear extension of the ROIs) of the optimal focus. Then, the second finest search of the focus range is repeated on evenly spaced points inside the targeted region. This procedure simultaneously optimizes the quality of the acquired images and the time spent on the autofocus search.
7. The entire ROI is acquired by collecting adjacent positions. Every channel is acquired sequentially.
8. In case of multi-ROIs selection, the same procedure is applied to each one.
9. The stored images are then analyzed by our in-house developed A.M.I.CO. analysis software.

Important The design of the sample acquisition can vary depending on the application. The acquisition modalities can be individually employed or sequentially iterated. We provide a general scheme for an experimental acquisition workflow in Fig. 1.

3.3 Image Analysis

The Automated Microscopy for Image CytOmetry (A.M.I.CO.) analysis package has been developed within the open-source ImageJ platform using its macro programming language. The software is freely available upon request or available on GitHub public repository (https://github.com/MarioFaretta/AMICO). The package is composed of two separated modules (A.M.I.CO_Union and A.M.I.CO_Plotting) and is described in detail in previous works [25]. In general, A.M.I.CO_Union focuses on data browsing, parameter setting, and image analysis processing, required for the recognition, segmentation, localization, and measurement of single cells. The second data-analysis module (A.M.I.CO_Plotting) provides an interface like flow cytometry analysis programs. Histograms and 2D multicolor dot plots are generated to represent all the events within the sample, enabling the definition of specific regions for statistical analysis and logical gates for the selective targeting of events. With the built-in cell retrieval procedure, each event (cell or subpopulation) can be visualized and physically relocated on the microscope stage for further operations (Fig. 2). The following procedure shows the main steps of A.M.I.CO. analysis:

3.3.1 Image Processing (A.M.I.CO_Union)

1. Download and install the A.M.I.CO. analysis package (by copying it into the plugins directory) in open-source ImageJ platform and run the program.
2. Open the first module, A.M.I.CO_Union.
3. Choose the folder directory containing the stored images.

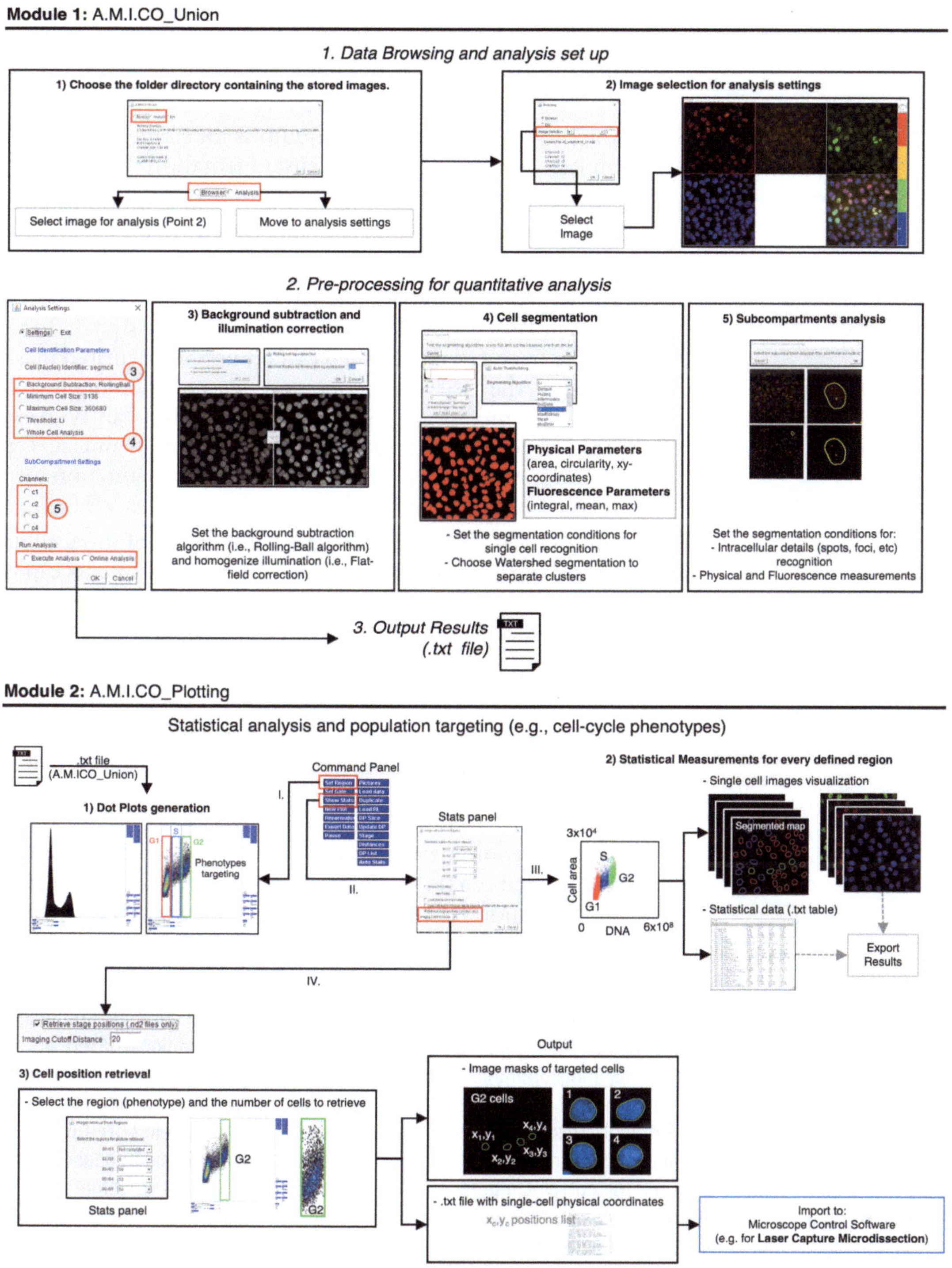

Fig. 2 Schematic representation of A.M.I.CO. computational platform. *Module 1 (A.M.I.CO_Union)*. The data browsing and analysis setup includes two main steps: the definition of the folder directory containing the images previously collected by multi-resolution acquisition workflow (**Step 1**) and a user-guided selection of the multichannel image for analysis setting (**Step 2**). The preprocessing procedure for quantitative analysis is

4. Set the required parameters for background subtraction by choosing the most suitable algorithm (i.e., rolling-ball algorithm) and to homogenize illumination (i.e., flatfield correction) (*see* **Note 4**).
5. Set the proper segmentation conditions for single-cell identification (i.e., Hoechst-based nuclei segmentation) (*see* **Note 5**).
6. Adjust the size parameters to define the smallest and largest areas that identify a cell, thus minimizing the potential production of cell fragments and aggregates.
7. Set the whole-cell analysis parameters if required. The software bases the identification of events on the strongest signal, that is, DNA-dye from cell nuclei, then extending the calculated masks to identify the cell cytoplasm and plasma membranes. In absence of a membrane marker, cell separation can be based on the Voronoi map created by the segmented nuclei. Cell extension calculation is then refined by segmentation of a cytoplasmic signal.
8. If required, set parameters for the recognition of intracellular structures (i.e., spots, foci, nucleoli). Subcompartments can be either nuclear or extend to the whole-cell space (*see* **Note 6**).
9. Set the output folder directory and run analysis.
10. Results are automatically saved at the end of the process (.txt file):

    After setting the proper segmentation conditions for single-cell identification, images are processed, and a single-cell data repository is created containing in each row single-cell measurements. Geometrical parameters (area, circularity, and physical XY-coordinates in the field-of-view) are calculated for each identified cell (in the form of a progressive number used as a tag). The fluorescence mean value per pixel is also computed,

Fig. 2 (continued) organized in these sequential steps: subtraction of background and correction of imaging artifacts (**Step 3**), cell identification (cell size definition and cell segmentation conditions) (**Step 4**), and subcompartments setting (optional) for measurements of intracellular details (**Step 5**). The image analysis creates a single-cell data repository in a.txt format with geometrical and fluorescence parameters (Output Results). *Module 2 (A.M.I.CO_Plotting)*. This module reads the output.txt file (from A.M.I.CO_Union) and creates a flow cytometry-like data analysis interface to target cell populations and statistical measurements (**Step 1**). In the "Command Panel" of dot plot interface, it is possible to define ROIs to target selected phenotypes by using "Set Region" option (I). "Show Stats" option opens the "Stats Panel" (II), allowing the statistical measurements for every defined region (III). For every region, it is possible to export single-cell images and statistical data (**Step 2**). Flagging the "retrieve stage position" option, every event of defined regions can be physically retrieved (IV). Physical cell location (baricentrum cell coordinates) are retrieved to select subpopulations of interest. The software generates images containing the masks of the targeted populations and a single-cell position list. The list could be imported into the microscope control to perform additional analysis (e.g., laser capture microdissection) (**Step 3**) together with the integrated fluorescence intensity per cell. For every analyzed channel, subcompartments can also be defined for the measurement of intracellular details (e.g., spots, nucleoli, etc.). Physical and fluorescence parameters are recalculated for each detected intracellular event. The analysis procedure stores data in a tab-delimited text file containing the calculated measurements, which are classified in cell- or subcompartment-related values, and the coordinate of the segmentation mask identifying single cells.

3.3.2 Data Analysis (A.M.I.CO_Plotting)

1. Open the second module, A.M.I.CO_Plotting.
2. Choose the directory of.txt file generated with A.M.I.CO_Union, containing single-cell data measurements.
3. The program generates an interface like flow cytometry analysis programs to calculate single-cells statistics.
4. Generate dot plots to represent all the events in the sample (*see* **Note 7**).
5. Define specific regions for statistical measurements and identify different phenotypes using regions for selecting target events. Use "Set Region" option to define the ROI in the dot plot.
6. Logical gates, corresponding to the union or intersection of multiple regions, can also be employed to restrict the analysis to a selected ensemble of events that satisfies quantitative criteria, excluding all the others from the calculation. Use "Set Gate" to define logic ensembles of regions. Clicking on the "Gates" button, it is possible to select the gate of interest on cells, subcompartments or both (*see* **Note 8**).
7. Select "Show Stats" option to create, for every region, single-cell images and.txt file containing statistical measurements of all parameters previously defined in the first module. For every region, it is possible to retrieve images of the corresponding cells, specifying the channel to be displayed.
8. Save.tif images and.txt files.
9. The "Show Stats" window shows "Cell Position Retrieval" option to physically repositioned selected cells (specific phenotypes) on the microscope stage (*see* **Note 9**).

3.3.3 Cell Position Retrieval

1. Open "Show Stats" window.
2. Select the region corresponding to a specific phenotype.
3. Choose the number of cells to be collected in the subsequent step.
4. Run analysis. The program generates a.txt file with the physical *xy*-coordinates (baricentrum cell coordinates) of gated cells (subpopulations of interest) (*see* **Note 10**).

5. The software creates images containing the masks of the targeted cells.
6. The.txt file containing the single-cell physical coordinates can now be imported in the microscope control software to relocate the target cells on microscope stage for laser capture microdissection or other downstream applications.

4 Notes

1. The protocol can also be adapted for Immuno-smRNA FISH to simultaneously detect proteins and RNA transcripts. In our lab, slides were prepared according to the manufacturer protocol: HCR RNA-FISH for FFPE Tissue Sections (Molecular Instruments Inc., Los Angeles, CA, USA).
2. Numerical aperture values of the objective span from 0.13 for very low magnification (4×) to as much as 1.49 for high-performance objectives (60X to 100X) that employ specialized immersion oils. High numerical aperture objectives collect more light and produce an image with better contrast. High signal-to-noise ratio increases the resolution near the theoretical limit.
3. The modern fluorescence microscope is the core of a wide range of technologies ranging from confocal to super resolution microscopy allowing, for example, to perform optical sectioning analysis to get a real 3D spatial distribution of the molecules in the tissue. This employed variable resolution workflow naturally introduces a "Multi-Modal Microscopy" approach allowing the switching between different imaging modalities, built around the microscope setup.
4. Background removal was crucial for the correct reconstruction of signal distribution. In absence of any background subtraction, the specific signal estimation (such as Hoechst or DAPI signal for DNA content estimation) is affected by aspecific fluorescence. A.M.I.CO. provides the following tools for background subtraction and illumination inhomogeneities removal: (i) simple background subtraction based on a fixed value (determined by a user-defined ROI), (ii) rolling-ball subtraction algorithm (included in ImageJ), and (iii) reference-based background subtraction coupled to flatfield correction. The rolling-ball algorithm estimates the background by calculating circles of size in the order of the objects present in the image. It is particularly useful when dealing with images that have uneven or nonuniform background intensity, such as in cells or tissues samples. The background is the part of the image that the circle of prefixed size can roll over without intersecting any

foreground objects. The efficiency of the rolling-ball algorithm may be partially compromised, if the radius of the reference circle is not correctly set, in presence of very high cell density, as neighboring cells could contribute to an overestimation of the background level. Once the background is estimated, it is subtracted from the original image, preserving the features of interest. Differently, reference-based background subtraction coupled to flatfield correction subtract the background and correct illumination disparities, caused by the imaging system. This approach employs a reference image (*Flatfield*), typically acquired by using fluorescent slides or thin fluorescent layers, to detect the illumination disparities. Background levels (*Background*) were instead inferred from mounted empty slices. Both the images (flatfield and background) were acquired separately for each channel. The corrected image was calculated according to the following procedure:

$$\text{Image}_{\text{corrected}} = \frac{\text{Image}_{\text{raw}} - \text{Background}}{\text{Flatfield} - \text{Background}}$$

5. Segmentation threshold can be computed according to the different methods available in the ImageJ software (http://fiji.sc/wiki/index.php/Auto_Threshold). Segmentation thresholding can be chosen depending on morphological and intensity values of the target (such as cells, particles, or intracellular features) and background. The Watershed segmentation algorithm provided by ImageJ is then applied to separate nuclei clusters. Artifacts created by nuclei over-segmentation were eliminated by the analysis of their size and circularity.
6. The segmentation analysis of subcompartments includes ad hoc image filters to enhance their detection, for example, Laplace-of-Gaussian filtering or Top-Hat transformation, based on the available ImageJ plugin.
7. Subcompartments can be considered as part of the cell of origin or as independent entities. Usually, physical and fluorescence parameters are calculated for each detected intracellular event, and summary values are assigned to the cell of origin (number of detected spots, average size per spot, total area of the cell covered by the spots, ratio between spots and cell area (fractional occupancy), mean fluorescence value per spot, total fluorescence related to the subcompartment, ratio between confined subcompartment, and total cell fluorescence for every analyzed channel). In the second case, statistics were calculated on the whole number of detected objects without any link to the originating cells.
8. Gates defined in each subcompartment can be applied to cells in different ways. In the standard mode, single objects in the subcompartment not satisfying the selection criteria defined by

the gate are excluded and subtracted from the total number of events per cell. If the option "Active on other subcompartments" is marked, only the cells containing at least one object in the gated subcompartment were included in the calculation.

9. Correct repositioning implies that the sample remained in the physical original location on the microscope. If, for some reason, the sample has been moved, the operation of cell retrieval fails, and the acquisition and analysis procedure must be restarted from scratch. Small systematic shift can be eventually corrected by calculating the drift and adding it to the retrieved coordinates.

10. A.M.I.CO generates the list of positions according to the xml formats of the Nikon control software. The structure of the list can be easily adapted to different formats required by other microscope control programs, if accessible.

Acknowledgments

S.P. is supported by the AIRC fellowship for Italy (Post-Doc ID 29538-2023). L.F. and M.F. research activity is supported by the Italian Ministry of Health with Ricerca Corrente and 5 × 1000 funds.

References

1. Bhatia S, Frangioni JV, Hoffman RM et al (2012) The challenges posed by cancer heterogeneity. Nat Biotechnol 30:604–610. https://doi.org/10.1038/nbt.2294
2. Marusyk A, Janiszewska M, Polyak K (2020) Intratumor heterogeneity: the Rosetta stone of therapy resistance. Cancer Cell 37:471–484. https://doi.org/10.1016/j.ccell.2020.03.007
3. Dornan D, Settleman J (2011) Dissecting cancer heterogeneity. Nat Biotechnol 29:1095–1096
4. Goltsev Y, Samusik N, Kennedy-Darling J et al (2018) Deep profiling of mouse splenic architecture with CODEX multiplexed imaging. Cell 174:968–981.e15. https://doi.org/10.1016/j.cell.2018.07.010
5. Metzker ML (2010) Sequencing technologies—the next generation. Nat Rev Genet 11:31–46. https://doi.org/10.1038/nrg2626
6. Alfaro JA, Bohländer P, Dai M et al (2021) The emerging landscape of single-molecule protein sequencing technologies. Nat Methods 18:604–617. https://doi.org/10.1038/s41592-021-01143-1
7. Goodwin S, McPherson JD, McCombie WR (2016) Coming of age: ten years of next-generation sequencing technologies. Nat Rev Genet 17:333–351. https://doi.org/10.1038/nrg.2016.49
8. Baslan T, Hicks J (2017) Unravelling biology and shifting paradigms in cancer with single-cell sequencing. Nat Rev Cancer 17:557–569. https://doi.org/10.1038/nrc.2017.58
9. Mund A, Coscia F, Kriston A et al (2022) Deep visual proteomics defines single-cell identity and heterogeneity. Nat Biotechnol 40:1231–1240. https://doi.org/10.1038/s41587-022-01302-5
10. Mansuri MS, Williams K, Nairn AC (2023) Uncovering biology by single-cell proteomics. Commun Biol 6:381. https://doi.org/10.1038/s42003-023-04635-2
11. Bennett HM, Stephenson W, Rose CM, Darmanis S (2023) Single-cell proteomics enabled by next-generation sequencing or mass spectrometry. Nat Methods 20:363–374. https://doi.org/10.1038/s41592-023-01791-5
12. Simone NL, Bonner RF, Gillespie JW et al (1998) Laser-capture microdissection: opening the microscopic frontier to molecular analysis.

Trends Genet 14:272–276. https://doi.org/10.1016/S0168-9525(98)01489-9

13. Espina V, Wulfkuhle JD, Calvert VS et al (2006) Laser-capture microdissection. Nat Protoc 1:586–603. https://doi.org/10.1038/nprot.2006.85
14. Cardoso dos Santos LM, Avila Y, Schvartz D et al (2023) Laser microdissection, proteomics, and multiplex immunohistochemistry: a bumpy ride into the study of paraffin-embedded fetal and pediatric lung tissues. Front Med 10:1191205
15. Martinek J, Wu T-C, Sun L et al (2022) Multiplex immunofluorescence-guided laser capture microdissection for spatial transcriptomics of metastatic melanoma tissues. STAR Protoc 3:101698. https://doi.org/10.1016/j.xpro.2022.101698
16. Ståhl PL, Salmén F, Vickovic S et al (2016) Visualization and analysis of gene expression in tissue sections by spatial transcriptomics. Science (80-) 353:78–82. https://doi.org/10.1126/science.aaf2403
17. Way GP, Sailem H, Shave S et al (2023) Evolution and impact of high content imaging. SLAS Discov 28:292–305. https://doi.org/10.1016/j.slasd.2023.08.009
18. Lin J-R, Chen Y-A, Campton D et al (2023) High-plex immunofluorescence imaging and traditional histology of the same tissue section for discovering image-based biomarkers. Nat Cancer 4:1036–1052. https://doi.org/10.1038/s43018-023-00576-1
19. Lucas AM, Ryder PV, Li B et al (2021) Open-source deep-learning software for bioimage segmentation. Mol Biol Cell 32:823–829. https://doi.org/10.1091/mbc.E20-10-0660
20. Schienstock D, Hor JL, Devi S, Mueller SN (2025) Cecelia: a multifunctional image analysis toolbox for decoding spatial cellular interactions and behaviour. Nat Commun 16:1931. https://doi.org/10.1038/s41467-025-57193-y
21. Hartmann J, Wong M, Gallo E, Gilmour D (2020) An image-based data-driven analysis of cellular architecture in a developing tissue. elife 9:e55913. https://doi.org/10.7554/eLife.55913
22. Furia L, Pelicci S, Perillo F et al (2022) Automated multimodal fluorescence microscopy for hyperplex spatial-proteomics: coupling microfluidic-based immunofluorescence to high resolution, high sensitivity, three-dimensional analysis of histological slides. Front Oncol 12:1–19. https://doi.org/10.3389/fonc.2022.960734
23. Furia L, Pelicci P, Faretta M (2014) High-resolution cytometry for high-content cell cycle analysis. Curr Protoc Cytom 70:7.41.1–7.41.15. https://doi.org/10.1002/0471142956.cy0741s70
24. Carpenter AE, Jones TR, Lamprecht MR et al (2006) CellProfiler: image analysis software for identifying and quantifying cell phenotypes. Genome Biol 7:R100. https://doi.org/10.1186/gb-2006-7-10-r100
25. Berg S, Kutra D, Kroeger T et al (2019) ilastik: interactive machine learning for (bio)image analysis. Nat Methods 16:1226–1232. https://doi.org/10.1038/s41592-019-0582-9
26. Bankhead P, Loughrey MB, Fernández JA et al (2017) QuPath: open source software for digital pathology image analysis. Sci Rep 7:16878. https://doi.org/10.1038/s41598-017-17204-5
27. Pelicci S, Furia L, Pelicci PG, Faretta M (2023) Correlative multi-modal microscopy: a novel pipeline for optimizing fluorescence microscopy resolutions in biological applications. Cells 12:354. https://doi.org/10.3390/cells12030354
28. Pelicci S, Furia L, Pelicci PG, Faretta M (2024) From cell populations to molecular complexes: multiplexed multimodal microscopy to explore p53-53BP1 molecular interaction. Int J Mol Sci 25:1–21. https://doi.org/10.3390/ijms25094672
29. Furia L, Pelicci PG, Faretta M (2013) A computational platform for robotized fluorescence microscopy (I): high-content image-based cell-cycle analysis. Cytometry A 83:333–343. https://doi.org/10.1002/cyto.a.22266

INDEX

Roberta Noberini (ed.), *Laser Capture Microdissection: Methods and Protocols*, Methods in Molecular Biology, vol. 3015, https://doi.org/10.1007/978-1-0716-5154-4,

MIX
Papier aus verantwortungsvollen Quellen
Paper from responsible sources
FSC® C105338

If you have any concerns about our products,
you can contact us on
ProductSafety@springernature.com

In case Publisher is established outside the EU,
the EU authorized representative is:
Springer Nature Customer Service Center GmbH
Europaplatz 3, 69115 Heidelberg, Germany

Printed by Libri Plureos GmbH
in Hamburg, Germany